THE USBORNE BOOK OF CUTAWAYS

CARS .... Page 1

PLANES .... Page 33

BOATS .... Page 65

INDEX .... Page 96

Edited by Cheryl Evans & Jane Chisholm

CUTAWAY CARS

Clive Gifford
Designed by Robert Walster

Additional illustrations by: Sean Wilkinson, John Scorey, Robert Walster

Usborne Publishing wish to thank the following for their help with this part of the book:

Citroën Ltd., Castrol International, Ford Motor Company Limited, Goodyear, Honda, Land Rover, Lotus, MIRA, Michelin Tyre PLC, Peugeot, Porsche Cars Great Britain Ltd., SAAB Automobile AB, Toyota, Volkswagen, Volvo

Contents

Early cars 2
Modern cars 4
Safety 6
Rally car 8
The engine 10
The transmission 12
Grand Prix racing car 14
Aerodynamics 16
Vintage racing car 18
Suspension and steering 20
Brakes and tyres 22
Sports car 24
Electrical system 26
Off-road car 28
The future 30
Glossary 31
Cars: Facts 32

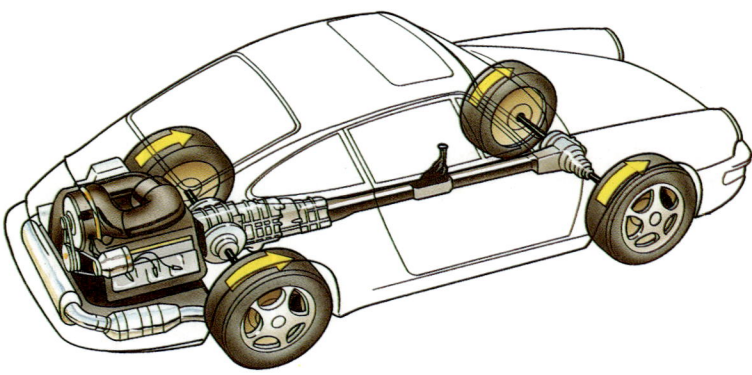

Words in *italic* type
Words in this section which appear in *italic* type and are followed by a small star (for example, *friction**) can be found in the glossary on page 31.

Early cars

The first vehicle to move on land without the help of a horse or other animal was Nicolas-Joseph Cugnot's steam tractor. Built in 1769, it could only run for 15 minutes at a time and its top speed was just 3.6km/h (2.2mph), slower than you can walk.

Cugnot's steam tractor

This is its steam boiler.

This luxury car is a Delaunay-Belleville F6. It was first built in 1908 and was popular with nobles and aristocrats in Europe.

This hood opens out to protect the passengers sitting in the back seats.

These wheels were originally designed for cannons. They are made of wood and have spokes like bicycle wheels.

Here you can see the back axle, a rod which joins the back wheels together.

Most of the car's body is made of wood and is either painted or varnished. Car bodies today are usually made of steel.

The first big step in making cars more like they are today came with the invention of a new type of engine powered by gas or petrol. It was called the internal combustion engine and you can learn more about it on pages 10-11.

This Benz Velo was built in 1898.

One of the first cars to be powered by this new engine was built by Karl Benz in 1885 (see page 32). Within ten years, his factory was building many cars for sale. One model, the Benz Velo, was the first car to sell in large numbers.

As more companies started to build cars, improvements such as proper brakes and lights for driving at night were added. More powerful engines combined with better car design made cars much faster and this resulted in many more accidents. Governments brought in laws about cars and speed for the first time.

In Britain, until 1896, a person waving a red flag as a warning had to walk in front of a car. This kept speeds down to under 6.5km/h (4mph).

More reliable

The first motor vehicles were not reliable and broke down all the time. As cars became more popular, car builders concentrated on improving them so that they ran better. Rolls Royce built their first luxury car, the Silver Ghost, in 1906. To demonstrate its reliability, a team of drivers drove it non-stop for 24,120km (14,988 miles). In all this time, the car only had to stop once for repairs.

This windshield folds down.

This engine is much larger but less than half as powerful as the engine in an ordinary family car today.

The headlight is powered by gas. Headlights on today's cars are powered by electricity.

This can is used to carry extra fuel. It is strapped firmly into place.

Most early cars need a strong turn of this starting handle to start their engine up. A modern car uses electricity to start its engine.

Delaunay-Belleville were famous for making steam boilers for trains and ships. In fact, the shape of this engine cover is rather like a steam boiler.

3

Modern cars

An ordinary Volkswagen Golf

One hundred years on from the first motor vehicles, modern cars look a lot different. Yet, the way they work is, in fact, very similar. For example, most cars still rely on an internal combustion engine to power them. Today's cars are more complicated than earlier models. They are made up of hundreds of parts all joined together to form what car engineers call systems.

A Golf used for rally racing

From design to production

It takes many years to design and build a modern car. First, the company researches what customers want and finds out what are the latest technical developments. They then start to choose some of the basic features they wish to include in the new car.

A car starts its life as drawings on a designer's desk. Changes are suggested by many people in the company. Everything from the seat colour to the size of the wheels is discussed.

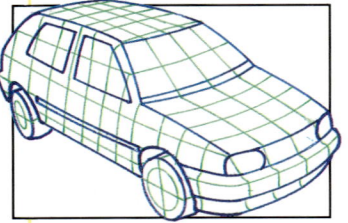

Engineers use Computer Aided Design (or C.A.D.) to determine the size and shape of the car and all its parts. Then, detailed plans and models of the car are made.

The models are tested in wind tunnels (see page 17) to see how they react to air moving over them. Many changes are made to the car's shape and testing lasts a long time.

Many other tests are done before the car can be produced in large numbers to sell to the public. Some of the most important testing is for safety (see pages 6-7).

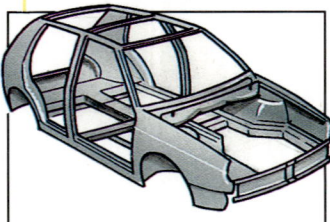

The parts of the car are made in several different factories. Some parts are even made by other companies. The car is then put together on a production line in a factory.

Eventually, cars come off the production line, are given a final test and are ready to be sold. The time between the original design and the first sales can be over five years.

Volkswagen Golf

The Volkswagen Golf is a popular family car. It is relatively small and compact but can carry up to five people and travel at speeds over 160km/h (100mph).

Wing mirror

This metal rod is the dipstick. It allows you to check how much oil is in the engine.

Engine

This is an air filter. It prevents dirt and dust from getting into the engine.

Headlight

This is the car's radiator. It helps cool the engine down.

Behind the radiator you can see the fan. This also helps to keep the engine cool.

Chassis and monocoque

The car's main parts used to be held in place by a frame called a *chassis**. Family cars today usually have a chassis combined with the body of the car. This is called a monocoque.

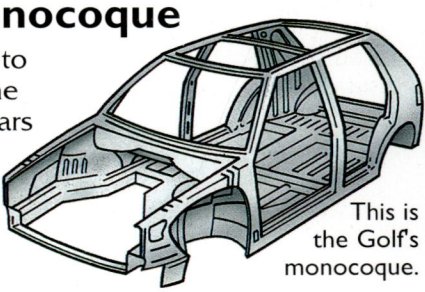

This is the Golf's monocoque.

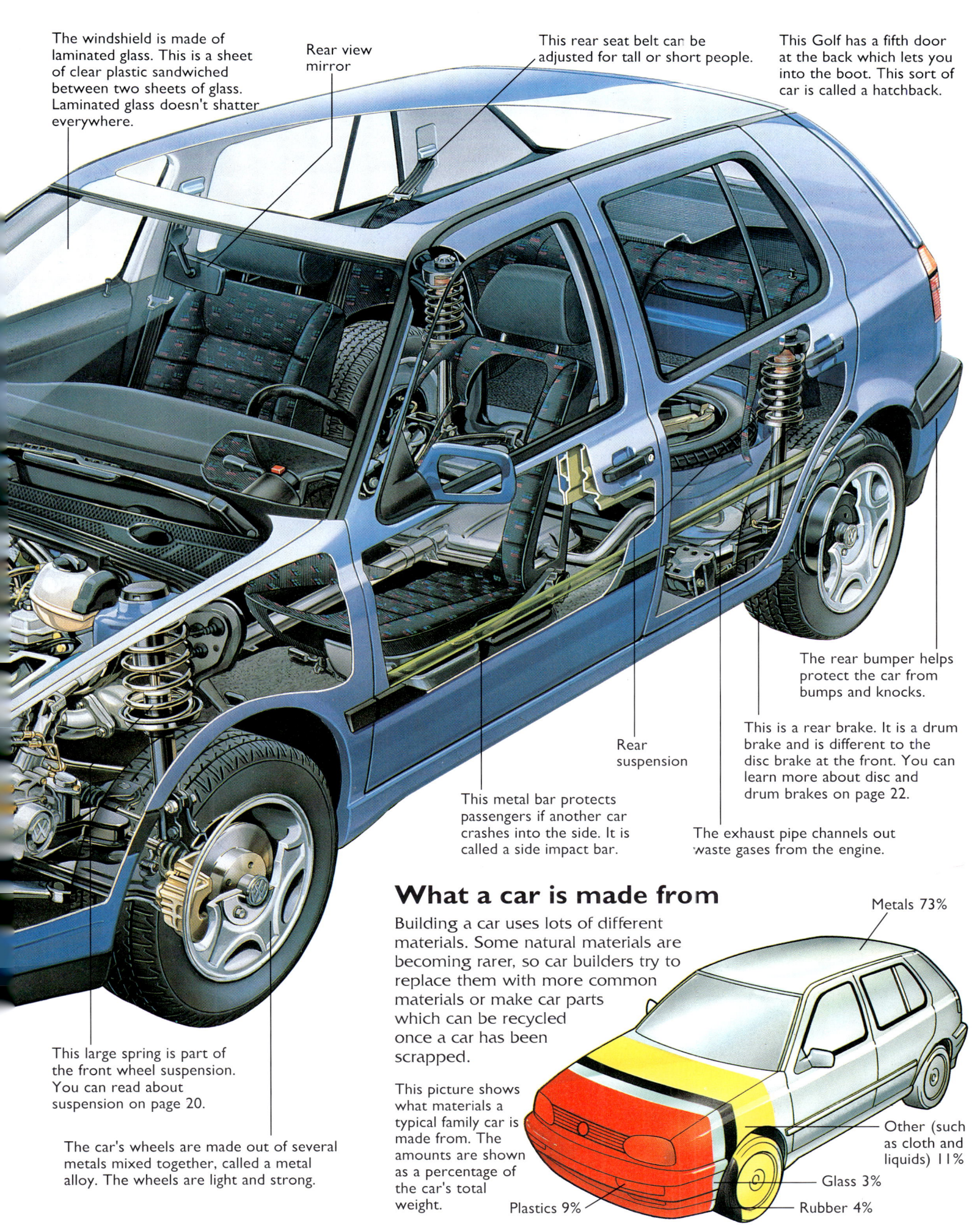

The windshield is made of laminated glass. This is a sheet of clear plastic sandwiched between two sheets of glass. Laminated glass doesn't shatter everywhere.

Rear view mirror

This rear seat belt can be adjusted for tall or short people.

This Golf has a fifth door at the back which lets you into the boot. This sort of car is called a hatchback.

The rear bumper helps protect the car from bumps and knocks.

Rear suspension

This is a rear brake. It is a drum brake and is different to the disc brake at the front. You can learn more about disc and drum brakes on page 22.

This metal bar protects passengers if another car crashes into the side. It is called a side impact bar.

The exhaust pipe channels out waste gases from the engine.

This large spring is part of the front wheel suspension. You can read about suspension on page 20.

The car's wheels are made out of several metals mixed together, called a metal alloy. The wheels are light and strong.

What a car is made from

Building a car uses lots of different materials. Some natural materials are becoming rarer, so car builders try to replace them with more common materials or make car parts which can be recycled once a car has been scrapped.

This picture shows what materials a typical family car is made from. The amounts are shown as a percentage of the car's total weight.

Metals 73%
Other (such as cloth and liquids) 11%
Glass 3%
Rubber 4%
Plastics 9%

Safety

Modern cars are designed to help the driver avoid accidents. The latest brakes, steering and tyres all give drivers more control of their car than ever before. These are called active safety features.

If there is an accident, a car's passive safety features protect the driver and passengers. The picture below shows some of the common passive safety features of a modern car in a crash test.

Dummies are used instead of people when a car is crash tested. Their movements and any damage they suffer is recorded using high speed photography and sensors linked to computers.

Crash test dummy

This is a steering wheel air bag (see how it works below).

To stop the driver's head from hitting the steering wheel, the steering column, to which the wheel is attached, can be made to collapse like a telescope.

This headrest stops the head from jolting sharply back. This action is called whiplash and can cause severe back and neck damage.

The seat is firmly fixed to the floor. It cannot slide back and trap the legs of a passenger in the back seat.

Seat belts hold people firmly in their seats. Many modern seat belts are fitted with powerful springs called pre-tensioners. They pull the belt tighter if there's a crash.

Submarining is when people are forced forward and under their seat belts by a crash. Modern seats are designed to stop this.

The front of the car body will crumple as the car hits something solid.

Steering wheel air bag

As a car hits something it starts to slow down and stop but the people inside the car keep moving. Many people in crashes are hurt by hitting their heads on the steering wheel or dashboard. An air bag should prevent this.

An air bag must inflate very quickly and stay blown-up until after the driver's head has hit it. This is done by igniting, or setting light to, chemicals which create large amounts of gases. These gases inflate the bag in an instant. Some cars have air bags to protect the front passenger as well.

Gases inflate air bag.

Inflating chemical is stored here.

Igniter sets light to chemicals to create gases.

Switches inside the car set off the chemicals in the air bag when the car crashes at over 33km/h (20mph).

The chemicals react and inflate the bag. A large cushion for the driver's chest and head is created.

The bag inflates in 40 milliseconds. That's less than a third of the time it takes you to blink your eye.

Body strength

Crashes create energy which has to go somewhere. Early car bodies were rigid. These protected against the direct impact of a crash only for the energy to travel through the car and throw people around.

A modern car still has a rigid body, called the passenger cell or cage, which will not break even if the car rolls over. Much of the rest of the car's body is designed to collapse when it is hit. The collapsing parts are called crumple zones and they absorb lots of energy from the crash. The remaining energy is directed around the car body but away from the driver and passengers.

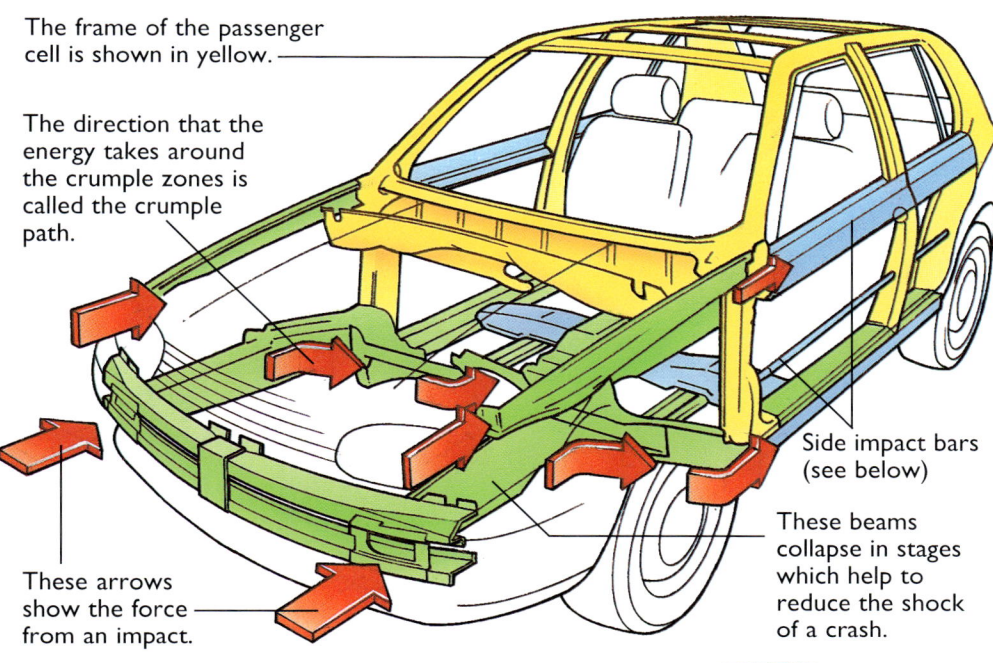

The frame of the passenger cell is shown in yellow.

The direction that the energy takes around the crumple zones is called the crumple path.

These arrows show the force from an impact.

Side impact bars (see below)

These beams collapse in stages which help to reduce the shock of a crash.

Side impact bars

Many accidents involve side-on crashes where one car punches a hole through another's door. Although some companies are starting to use side air bags, the most common way to protect people is to put strong rods of steel, called side impact bars, inside the door frame.

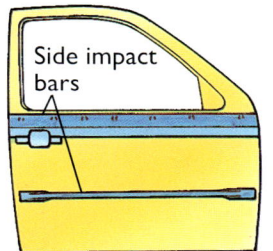

Side impact bars

Computer simulation

Some car companies use powerful computers to improve car safety. They build a model of the car in the computer which is accurate right down to the last detail. Although this takes a long time to do, it allows every part of the car to be crash tested in the computer. Engineers can alter the size or the strength of a piece of the car at the touch of a few buttons, and see the effect more quickly than building a proper part which has to be checked in a real car crash test. (You can find out about flight simulation on page 49.)

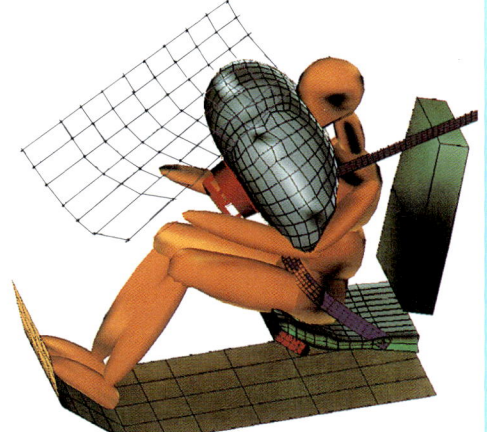

This is a computer test of the air bag opening.

The tests can take as long as 30 hours to complete but provide the safety team with lots of important information.

Crash testing

Crash tests have to cover all the different possible types of crash. Engineers record what happens to a car in each crash by using photography and electronic instruments, both inside and outside of the car. They can then tell what changes may be needed to improve the car's design.

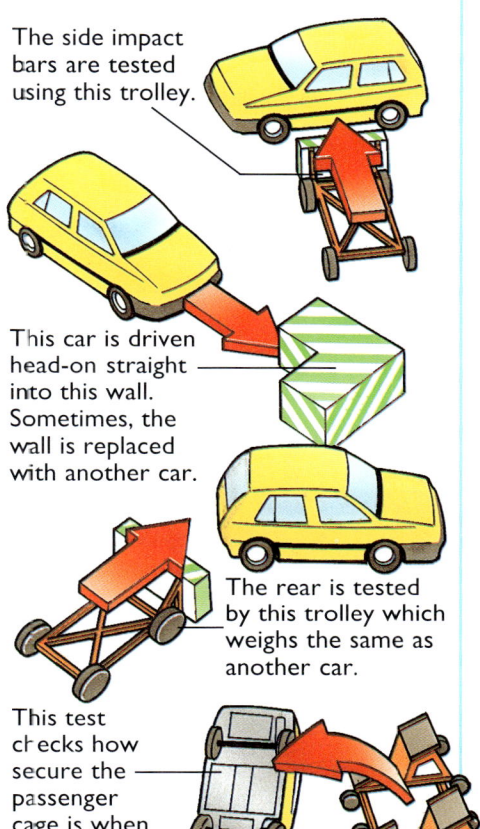

The side impact bars are tested using this trolley.

This car is driven head-on straight into this wall. Sometimes, the wall is replaced with another car.

The rear is tested by this trolley which weighs the same as another car.

This test checks how secure the passenger cage is when the car rolls over.

This is an overhead view of a crash test to the front of a car.

Rally car

Rally cars race in difficult conditions: through deserts, over icy roads and on bumpy dirt tracks, for example. They have to be tough to take the battering they receive. Most rally cars are specially modified versions of normal family cars.

Ford Escort RS Cosworth

This car has won a lot of competitions. It is based on a high speed version of the Ford Escort. It has many features added including a strengthened body, tougher, lighter wheels and a bigger, more powerful engine.

This is a standard Ford Escort.

Wings

This large flap is called a wing. It helps the car grip the road and makes it easier to drive. You can learn more about how it works on page 14.

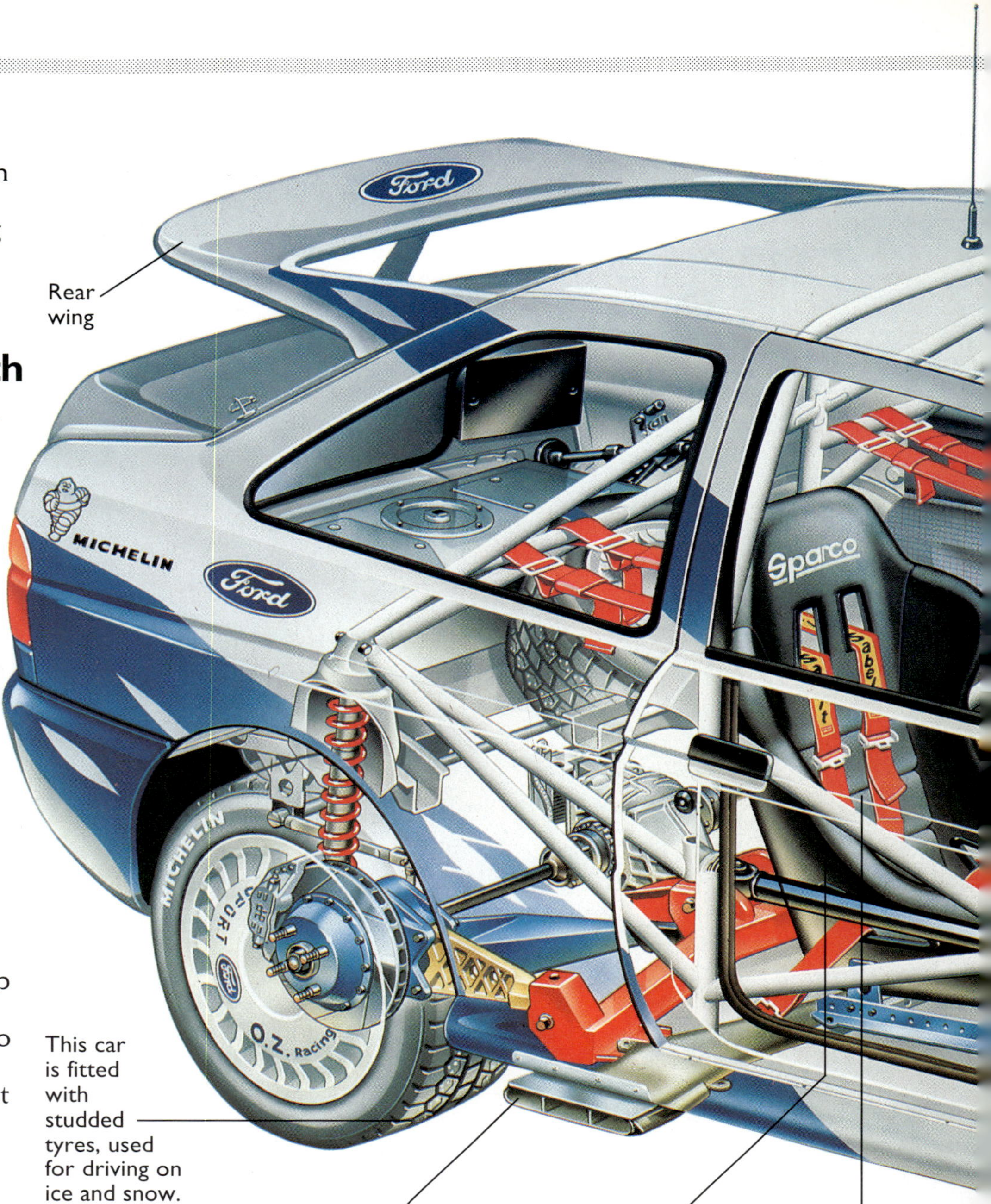

Rear wing

This car is fitted with studded tyres, used for driving on ice and snow.

This is the car's exhaust pipe. Its unusual flat shape stops it from hitting bumps in the ground.

These seats are specially shaped to hold the driver and co-driver. They are called bucket seats.

The co-driver sits here. He or she plots the car's route around the rally course.

Driver protection

Rally cars crash and roll over more often than family cars. In a rally car, the driver and co-driver are protected by a strong frame of steel tubes fitted to the inside of the car body. This frame is called a roll cage.

The driver and co-driver are held firmly in their seats by a set of five straps called a racing harness. These straps go around the waist, over the shoulders and between the legs. They all fasten together at the front.

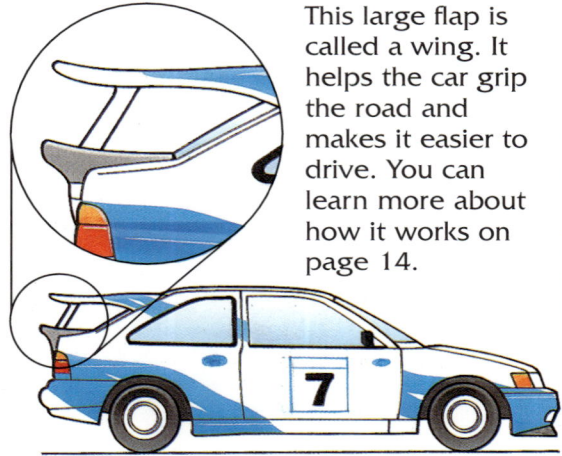

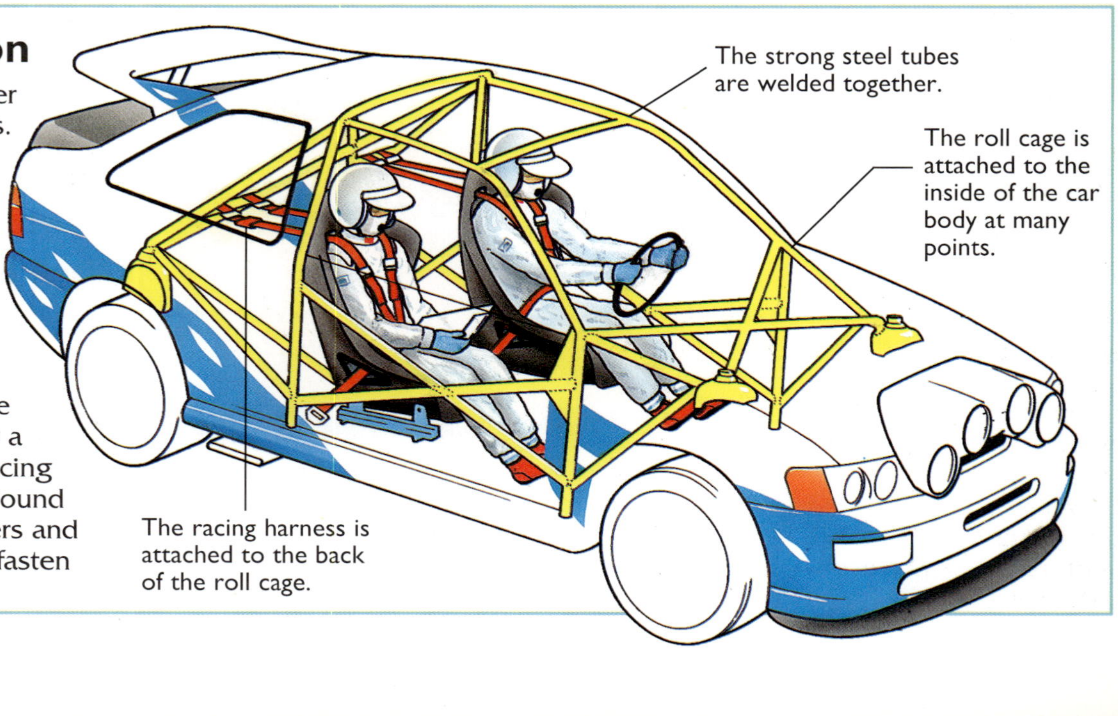

The strong steel tubes are welded together.

The roll cage is attached to the inside of the car body at many points.

The racing harness is attached to the back of the roll cage.

Lights

The Escort Cosworth sometimes has four extra front lights for rallying at night or racing along tracks in dark forests. Each of these lights is about twice as strong as a normal car headlight.

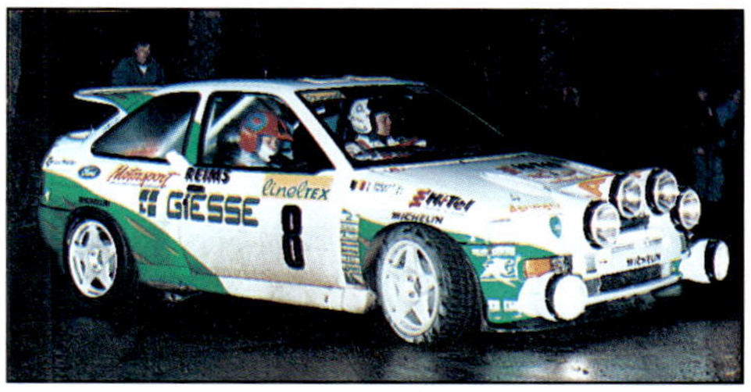

This Ford Escort RS Cosworth is rallying at night with all of its lights on.

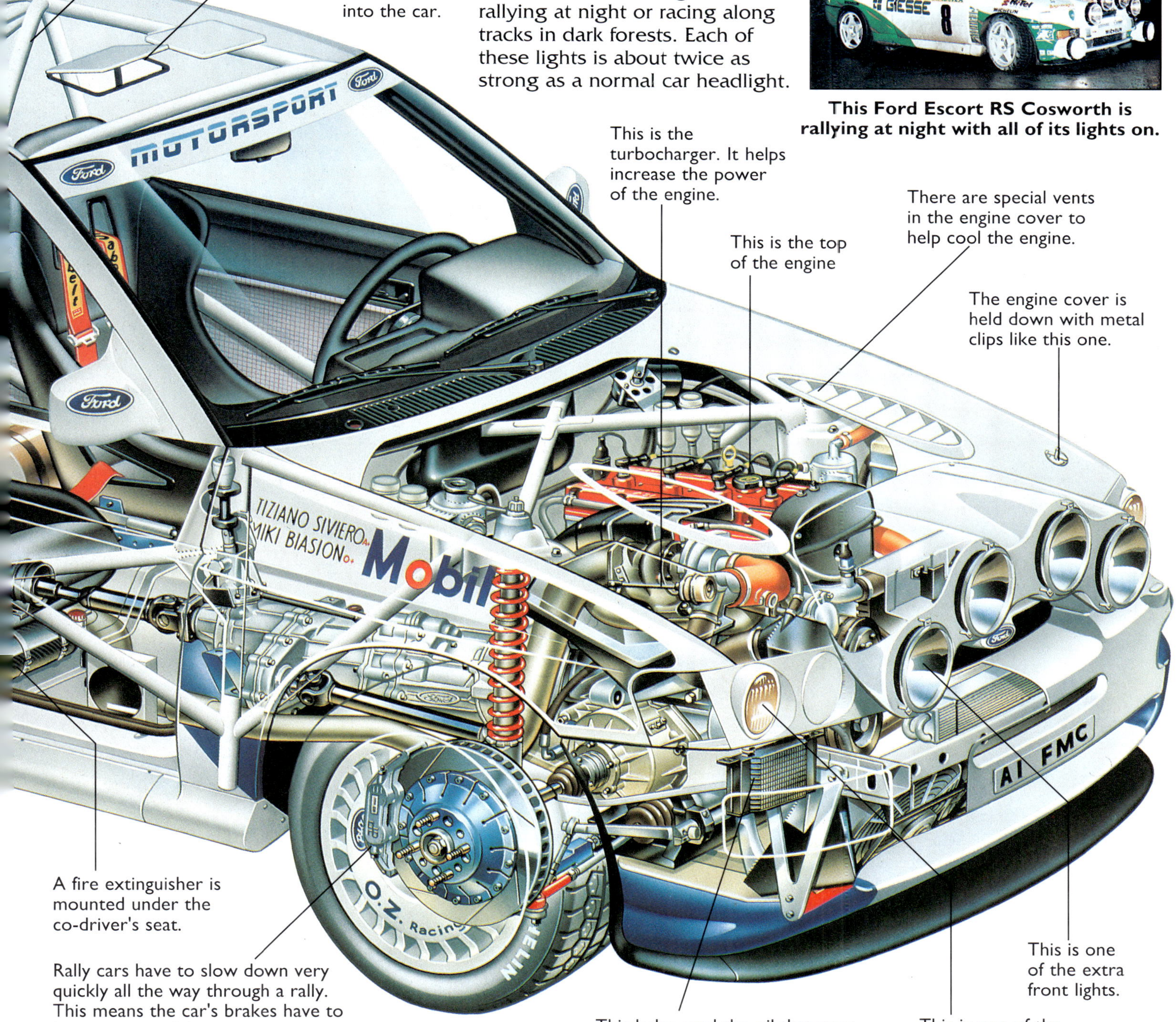
The body is cut away here to show part of the roll cage (see below).
This roof vent lets air into the car.
This is the turbocharger. It helps increase the power of the engine.
This is the top of the engine
There are special vents in the engine cover to help cool the engine.
The engine cover is held down with metal clips like this one.
A fire extinguisher is mounted under the co-driver's seat.
Rally cars have to slow down very quickly all the way through a rally. This means the car's brakes have to be very powerful.
This helps cool the oil that runs around the engine.
This is one of the ordinary headlights.
This is one of the extra front lights.
MOTORSPORT
TIZIANO SIVIERO
MIKI BIASION
Mobil
A1 FMC

Rally racing

Rally courses are a mixture of roads and dirt tracks. The course is divided into separate sections known as Special Stages. One stage may be along twisting and turning mountain roads, another on mud tracks through dark forests. The cars follow the same route but start one after another. Each car is timed over the different stages by officials called marshals. The winner is the car which has the fastest overall time.

Cars at the bigger rallies are supported by teams of dozens of people and many vehicles. These can include motorcycles to carry messages, trucks holding spare parts, a medical van and, sometimes, even a helicopter.

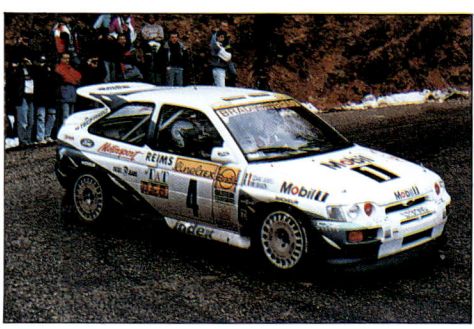

This car, without extra lights, is racing along a twisting mountain road.

The engine

Most cars are powered by a type of engine called an internal combustion, or I.C., engine. It is called this because it produces power by combusting, or burning, a mixture of fuel and air inside a chamber called a cylinder.

How an I.C. engine works

Here are the names of many of the important parts at the heart of an I.C. engine.

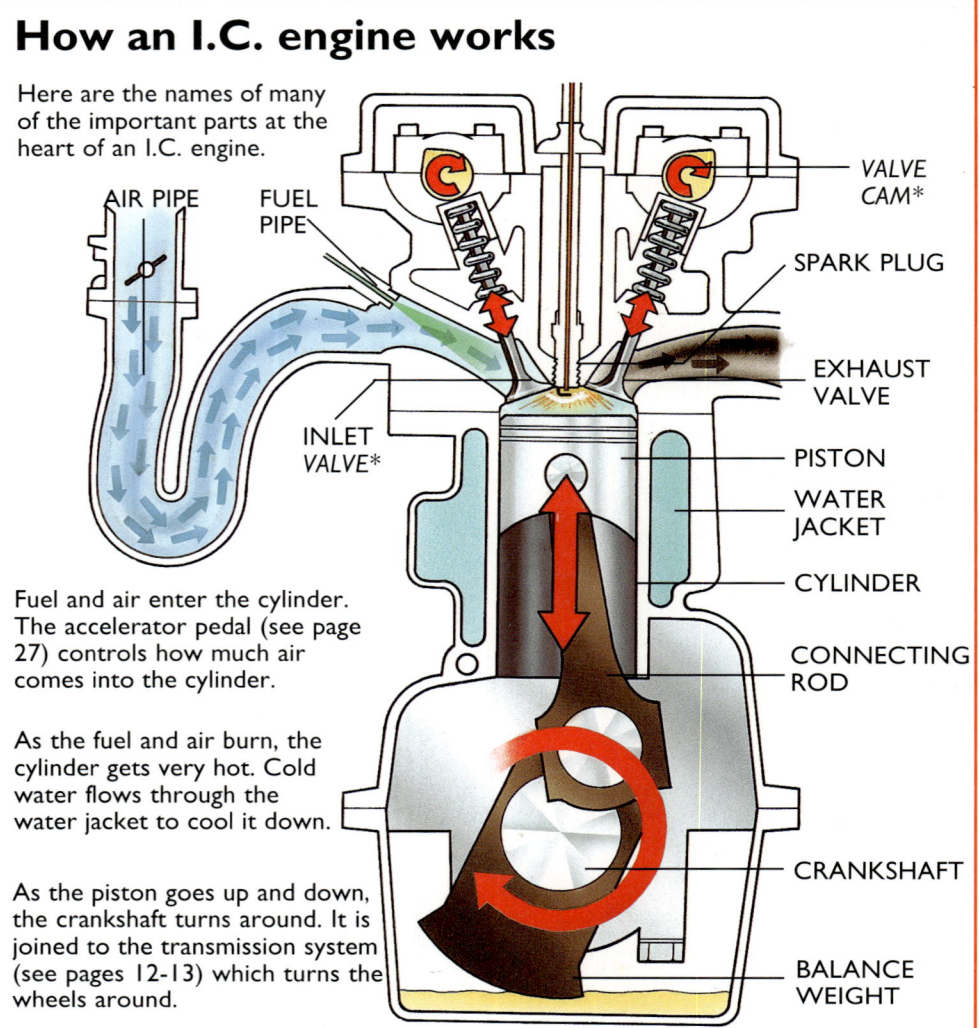

Fuel and air enter the cylinder. The accelerator pedal (see page 27) controls how much air comes into the cylinder.

As the fuel and air burn, the cylinder gets very hot. Cold water flows through the water jacket to cool it down.

As the piston goes up and down, the crankshaft turns around. It is joined to the transmission system (see pages 12-13) which turns the wheels around.

The actions that create power in the cylinder are called the combustion cycle. Most car engines have a cycle of four strokes. This means that the piston moves up twice and down twice in one complete cycle.

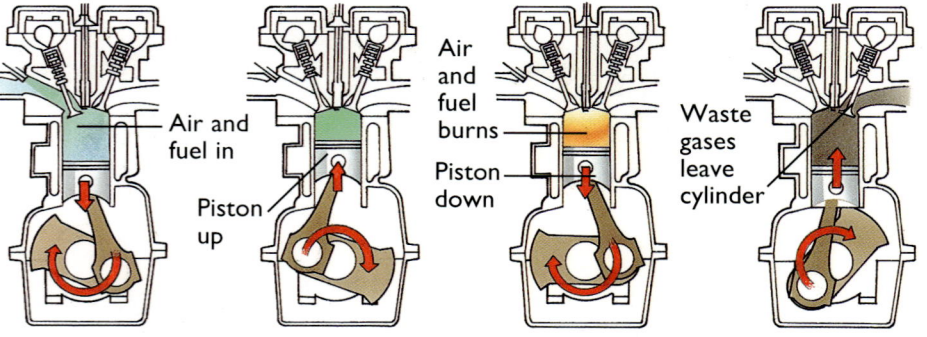

1. At the cycle's start, the piston moves down and the inlet valve opens. The fuel and air mixture is sucked into the cylinder.

2. The piston moves up the cylinder. This compresses and heats up the fuel and air. The spark plug sets light to the mixture.

3. The mixture burns, creating gases which expand quickly. These push the piston down. This stroke produces the engine's power.

4. The exhaust valve opens and the waste gases are pushed out of the cylinder by the rising piston. The engine then starts another complete cycle.

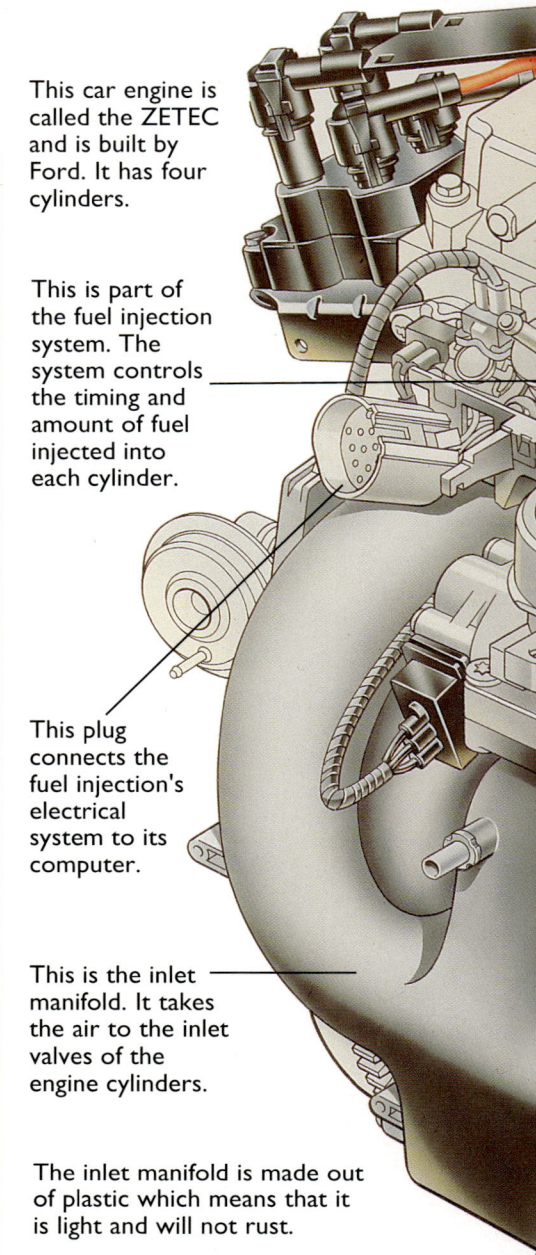

This car engine is called the ZETEC and is built by Ford. It has four cylinders.

This is part of the fuel injection system. The system controls the timing and amount of fuel injected into each cylinder.

This plug connects the fuel injection's electrical system to its computer.

This is the inlet manifold. It takes the air to the inlet valves of the engine cylinders.

The inlet manifold is made out of plastic which means that it is light and will not rust.

How many cylinders?

A modern car engine has more than one cylinder. A small engine may have four while engines used in powerful racing or sports cars can have as many as twelve. They can be arranged in different patterns.

This engine's four cylinders are arranged in line.

This engine's eight cylinders are arranged in a V pattern.

This combination is called a flat six.

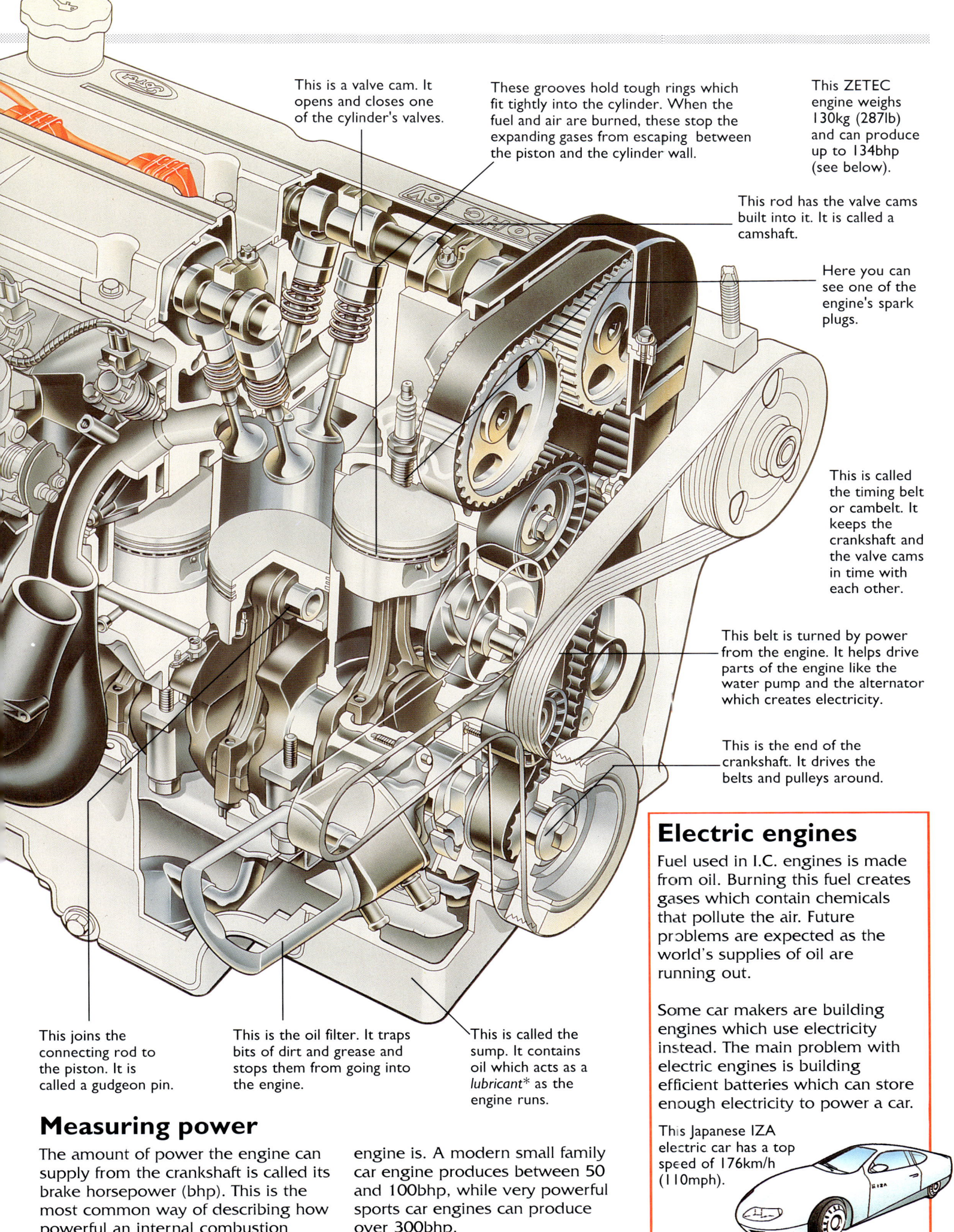

This is a valve cam. It opens and closes one of the cylinder's valves.

These grooves hold tough rings which fit tightly into the cylinder. When the fuel and air are burned, these stop the expanding gases from escaping between the piston and the cylinder wall.

This ZETEC engine weighs 130kg (287lb) and can produce up to 134bhp (see below).

This rod has the valve cams built into it. It is called a camshaft.

Here you can see one of the engine's spark plugs.

This is called the timing belt or cambelt. It keeps the crankshaft and the valve cams in time with each other.

This belt is turned by power from the engine. It helps drive parts of the engine like the water pump and the alternator which creates electricity.

This is the end of the crankshaft. It drives the belts and pulleys around.

This joins the connecting rod to the piston. It is called a gudgeon pin.

This is the oil filter. It traps bits of dirt and grease and stops them from going into the engine.

This is called the sump. It contains oil which acts as a *lubricant** as the engine runs.

Measuring power

The amount of power the engine can supply from the crankshaft is called its brake horsepower (bhp). This is the most common way of describing how powerful an internal combustion engine is. A modern small family car engine produces between 50 and 100bhp, while very powerful sports car engines can produce over 300bhp.

Electric engines

Fuel used in I.C. engines is made from oil. Burning this fuel creates gases which contain chemicals that pollute the air. Future problems are expected as the world's supplies of oil are running out.

Some car makers are building engines which use electricity instead. The main problem with electric engines is building efficient batteries which can store enough electricity to power a car.

This Japanese IZA electric car has a top speed of 176km/h (110mph).

11

The transmission

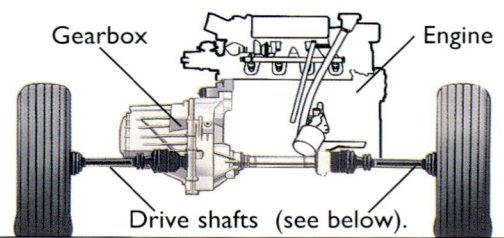
Gearbox — Engine — Drive shafts (see below).

The transmission system sends power from the engine to the wheels. The first transmissions were like bicycle chains, but in today's cars they are made up of many parts. A modern transmission passes the power from the engine through a gearbox, which gives the driver a chance to select different speeds.

The power is then taken to whichever wheels push the car forward. Twenty years ago, most cars were powered by their rear wheels but today more and more cars have a transmission system which drives the front wheel like the one shown below. Some cars are even driven by all four wheels (see page 24).

Gearbox

The heart of a transmission system is the gearbox. This system, made by Saab, has a gearbox with five forward gears and one reverse. Here you can see the names of some of its parts. The explanations on these two pages tell you how it all works.

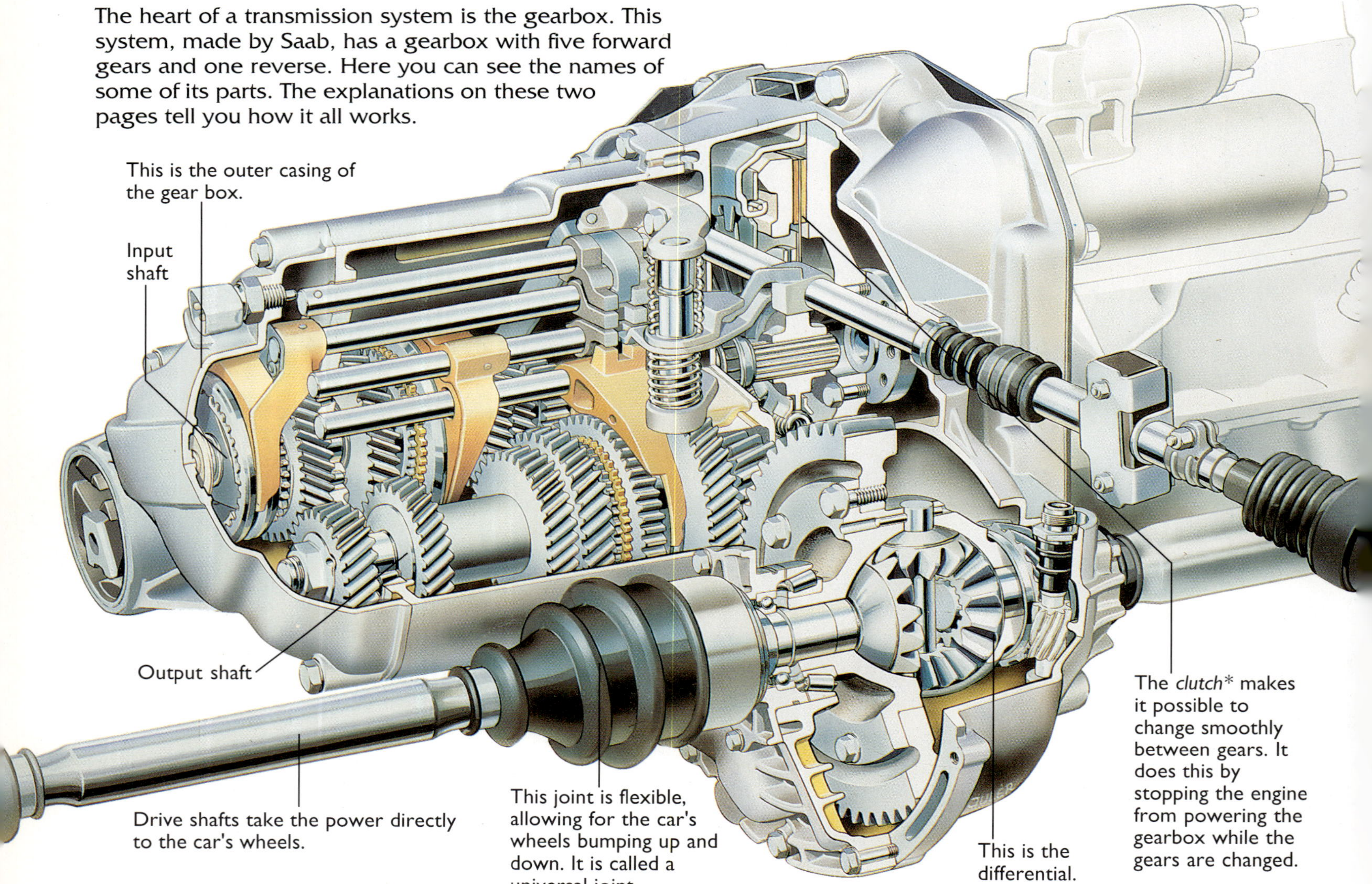

This is the outer casing of the gear box.

Input shaft

Output shaft

Drive shafts take the power directly to the car's wheels.

This joint is flexible, allowing for the car's wheels bumping up and down. It is called a universal joint.

This is the differential.

The *clutch** makes it possible to change smoothly between gears. It does this by stopping the engine from powering the gearbox while the gears are changed.

What are gears?

The gears found in a car gearbox are called cogs and are like wheels with teeth. The teeth allow gears to interlock, or mesh. When the gears mesh together, turning one gear around makes the other turn, but in the other direction.

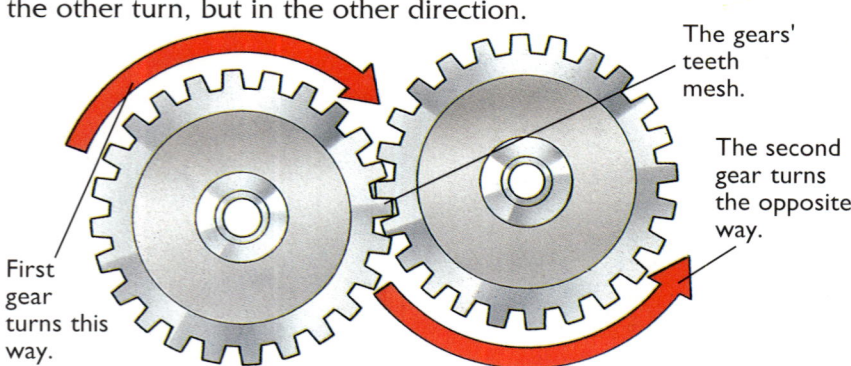

The gears' teeth mesh.

The second gear turns the opposite way.

First gear turns this way.

The small gear turns four times for each turn of the large.

Large gear four times bigger than small gear.

Large gear turns with four times more torque (see below) than the small gear.

If the gears are different sizes, the smaller one turns faster but the slower, bigger gear turns with greater power. Turning power is called *torque** by engineers and mechanics.

How the gearbox works

A car's gearbox contains many gear cogs which together provide the car with four, five or six different speeds. A car needs these because driving requires different combinations of speed and force at different times. For example, driving up a steep hill needs more force than speed while cruising along a motorway needs more speed than force. The lower gears in a gearbox provide greater force and the higher gears more speed.

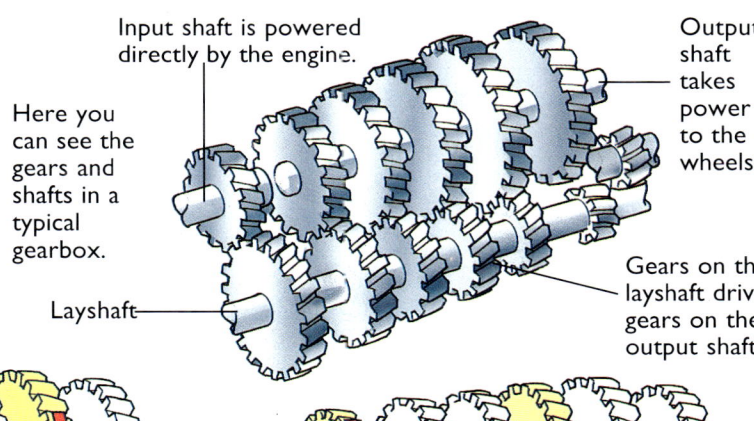

Input shaft is powered directly by the engine.

Output shaft takes power to the wheels.

Here you can see the gears and shafts in a typical gearbox.

Layshaft

Gears on the layshaft drive gears on the output shaft.

On top of the gear stick is this pattern showing where each gear is.

Gear stick

Gear shift rod

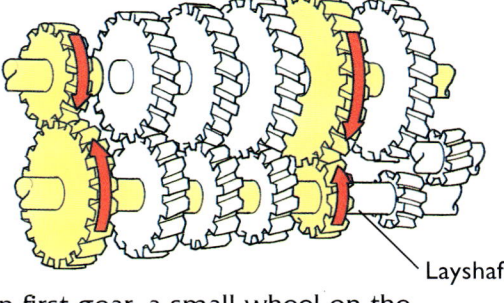

Layshaft

In first gear, a small wheel on the layshaft and a large wheel on the output shaft connect. The car moves slowly but with much power.

In second gear, the two gear cogs have less difference in size. This gear turns more quickly but with less power. This helps the car pull away with ease.

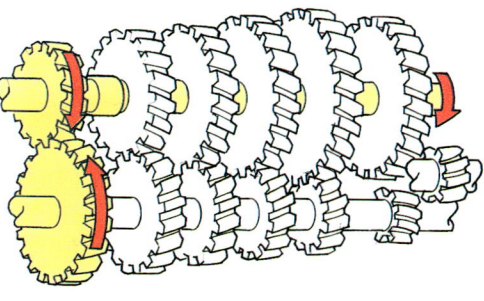

In the highest gear, the cogs on the input and output shafts are linked so that they turn at the same rate. This top gear allows the car to go at its fastest.

Output shaft turns opposite way to usual.

Idler gear

In reverse, an extra gear, called an idler gear, slips in between the normal input and output gears. The output shaft and the car's wheels turn the opposite way.

What is the differential?

The differential is a complicated set of gears between the two drive shafts of whichever wheels power the car. The differential adjusts the speed at which the two wheels turn when the car goes around a corner. The outer wheel has to travel farther around a corner than the inner one. The differential's gears make the outer wheel turn faster than the inner one.

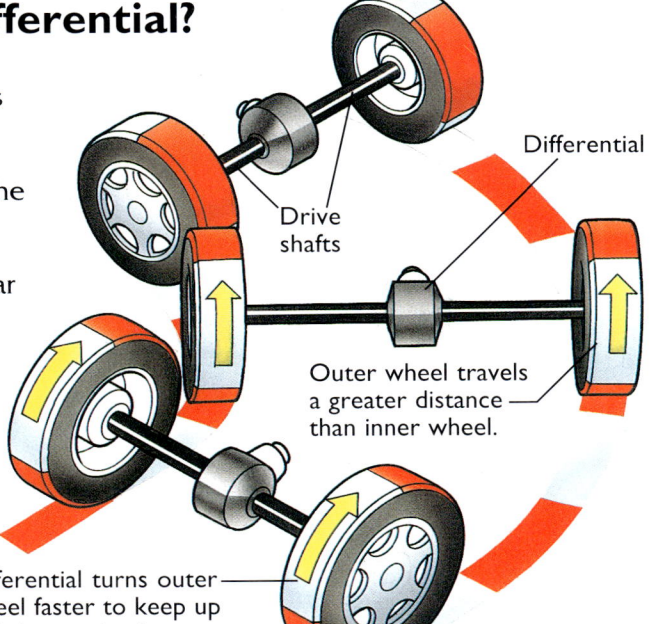

Differential

Drive shafts

Outer wheel travels a greater distance than inner wheel.

Differential turns outer wheel faster to keep up with inner wheel.

Automatic gearbox

Changing from one gear to another constantly can be tiring. One solution is an automatic gearbox that selects the engine gears for you. Automatic transmissions tend to use more fuel and the driver still has to make some choices by selecting with the gear stick.

P is for when the car is parked.

D is for drive, the main forward gear.

R is reverse gear.

13

Grand Prix racing car

Grand Prix racing is the most famous type of racing. With their powerful engines and light, specially-designed bodies, Grand Prix cars can travel at speeds of up to 320km/h (200mph). Grand Prix cars are often the first to use new technology. Some of the advanced features which prove useful for normal road drivers are eventually seen on ordinary family cars.

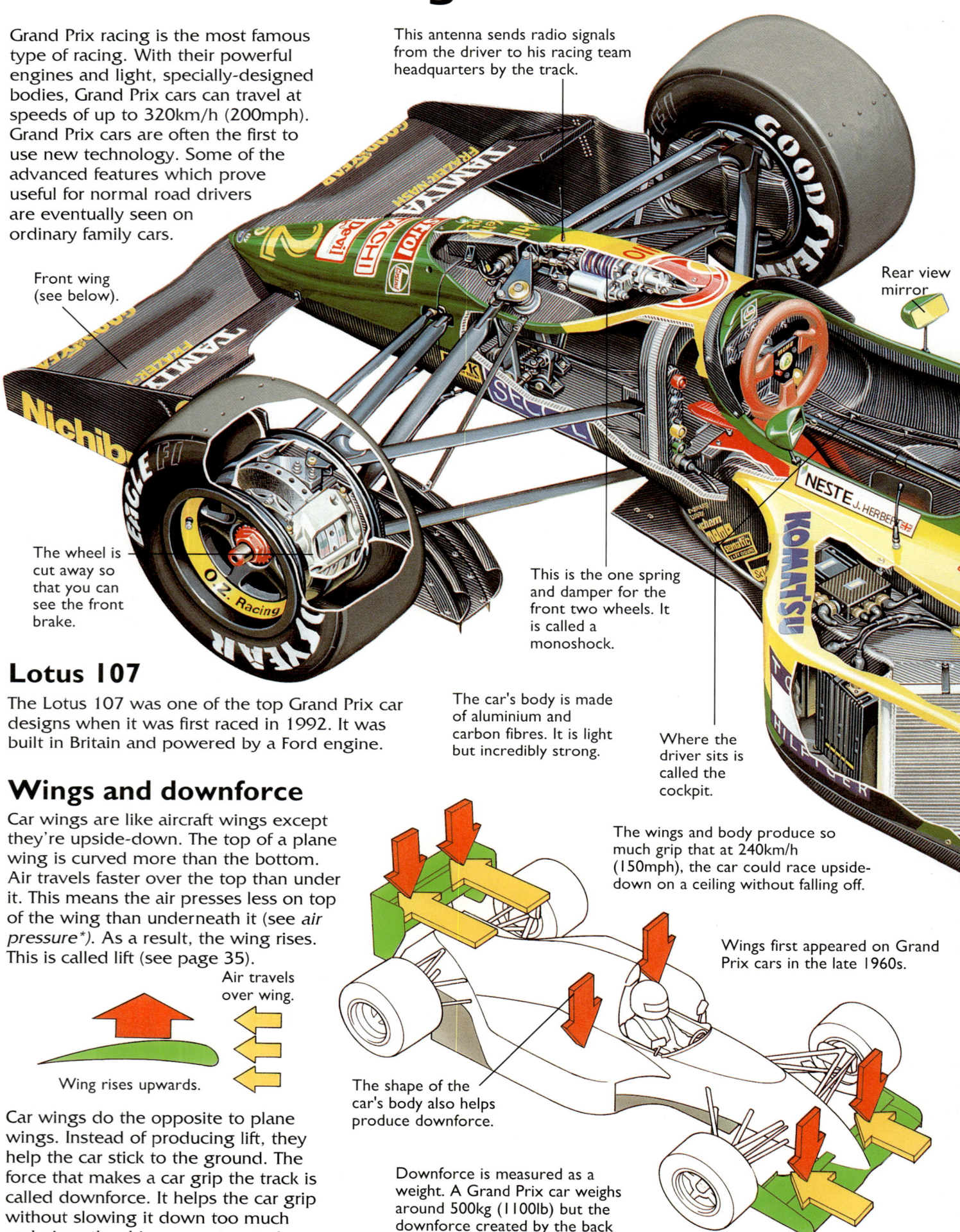

This antenna sends radio signals from the driver to his racing team headquarters by the track.

Front wing (see below).

Rear view mirror

The wheel is cut away so that you can see the front brake.

This is the one spring and damper for the front two wheels. It is called a monoshock.

The car's body is made of aluminium and carbon fibres. It is light but incredibly strong.

Where the driver sits is called the cockpit.

The wings and body produce so much grip that at 240km/h (150mph), the car could race upside-down on a ceiling without falling off.

Wings first appeared on Grand Prix cars in the late 1960s.

The shape of the car's body also helps produce downforce.

Downforce is measured as a weight. A Grand Prix car weighs around 500kg (1100lb) but the downforce created by the back and front wings can be as much as three times that.

Lotus 107

The Lotus 107 was one of the top Grand Prix car designs when it was first raced in 1992. It was built in Britain and powered by a Ford engine.

Wings and downforce

Car wings are like aircraft wings except they're upside-down. The top of a plane wing is curved more than the bottom. Air travels faster over the top than under it. This means the air presses less on top of the wing than underneath it (see *air pressure**). As a result, the wing rises. This is called lift (see page 35).

Air travels over wing.

Wing rises upwards.

Car wings do the opposite to plane wings. Instead of producing lift, they help the car stick to the ground. The force that makes a car grip the track is called downforce. It helps the car grip without slowing it down too much and gives the driver more control when turning.

Grand Prix engines

Grand Prix racing pushes a car, and especially the engine, to its absolute limits. Most engines are completely rebuilt after each race. The mechanics study the telemetrics, which are the performance details of the car recorded on computer during the race. They then rebuild and modify the engine according to the results.

This is the Lotus 107 raced by British driver, Johnny Herbert.

This Ford Cosworth HB engine powers the Lotus 107 and is protected by an advanced lubricant made by its sponsor, Castrol.

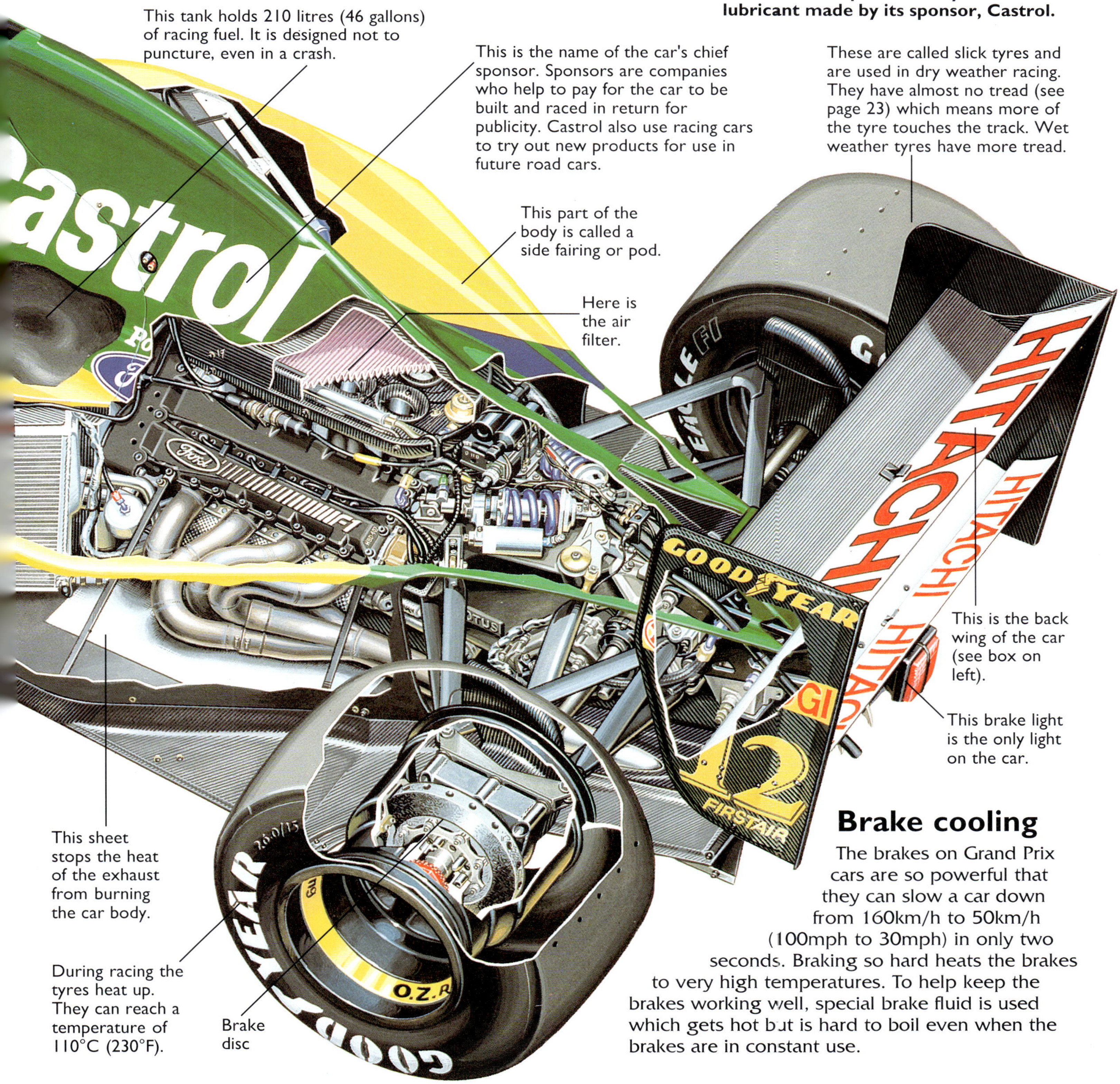

This tank holds 210 litres (46 gallons) of racing fuel. It is designed not to puncture, even in a crash.

This is the name of the car's chief sponsor. Sponsors are companies who help to pay for the car to be built and raced in return for publicity. Castrol also use racing cars to try out new products for use in future road cars.

These are called slick tyres and are used in dry weather racing. They have almost no tread (see page 23) which means more of the tyre touches the track. Wet weather tyres have more tread.

This part of the body is called a side fairing or pod.

Here is the air filter.

This is the back wing of the car (see box on left).

This brake light is the only light on the car.

This sheet stops the heat of the exhaust from burning the car body.

During racing the tyres heat up. They can reach a temperature of 110°C (230°F).

Brake disc

Brake cooling

The brakes on Grand Prix cars are so powerful that they can slow a car down from 160km/h to 50km/h (100mph to 30mph) in only two seconds. Braking so hard heats the brakes to very high temperatures. To help keep the brakes working well, special brake fluid is used which gets hot but is hard to boil even when the brakes are in constant use.

15

Aerodynamics

To understand aerodynamics, you need to know about *friction**. Friction is the scientific name for when two things rub together. You can feel the effects of friction if you rub your hands together.

The tighter you press your hands together the more effort it takes to move them. After a short while, your hands start to warm up. If you kept rubbing them for a long time, the friction would cause wear and you would get blisters.

Friction wastes power, creates heat and, over time, wears down the surfaces of objects rubbing together. Reducing friction means that cars can move faster and with less effort.

Without friction, a car wouldn't go at all. Friction between wheels and the ground allows the wheels to grip the ground and push the car forward. A car's brakes also rely on friction to slow the wheels down.

The tyres grip even when the car moves up a steep slope.

The friction of air

When air and a moving object rub together they create friction. Today's cars are designed to cause less air friction than cars in the past. Their particular shapes and some of their features were developed by the use of aerodynamics.

This is a typical car of the 1920s.

Box shape
Sharp corners

Air hits the front of the car flat on. The air cannot easily pass over and around the car so it creates lots of air friction.

Aerodynamics is the study of how a moving object travels through the air. It was first developed to look at how aircraft fly but is now used on motor vehicles as well.

Modern cars have smooth, rounded shapes (known as *streamlined**) which air passes over easily. This reduces the amount of friction from the air.

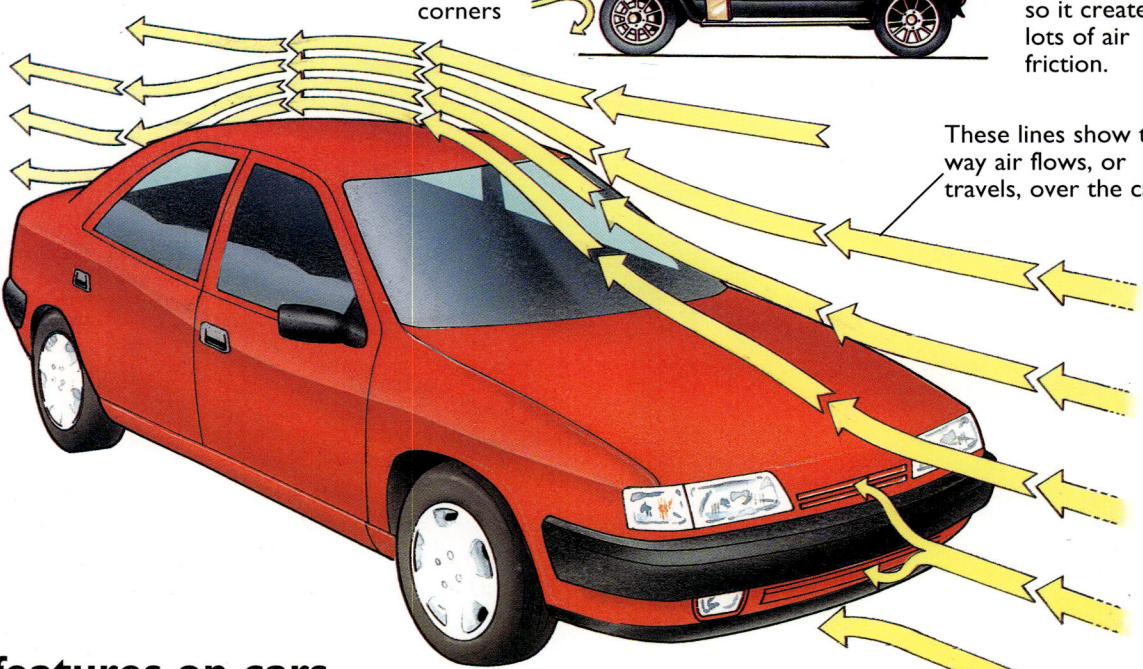

These lines show the way air flows, or travels, over the car.

Streamlined features on cars

Wheel arches fit closely around the wheel.

Door handles fit flush into the car body.

The lights are moulded into the body of the car.

The windshield is angled back.

The wing mirrors are rounded and smooth.

Drag coefficient

The drag coefficient, or Cd, is a measure of how much air friction a particular car will encounter. Less air friction means the car uses less fuel. Drag coefficients have got smaller as cars have become more streamlined. A family car in the 1960s had a drag coefficient of around 0.5 or 0.45. In today's family car, it is around 0.3.

This Vesta 2 research car, built by Renault, has a drag coefficient of only 0.19.

16

Reducing friction

In a car, the surfaces of moving parts that rub together are made as smooth as possible to reduce friction. A liquid is far smoother than the surfaces of the car parts. A thin layer of liquid, usually oil, placed between the moving parts will produce a smoother surface and less friction. A liquid used in this way is called a *lubricant**.

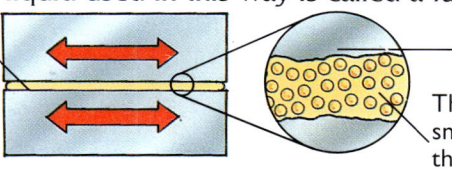

Thin layer of oil between two moving surfaces.

Metal

The oil is much smoother than the metal.

A rolling movement creates less friction than a sliding one. You can see this for yourself. First slide a book across a table. Then put some marbles underneath it and push it across the table again. The marbles roll rather than slide and this rolling creates less friction which means the book moves more easily. In some car parts, steel balls do the same job as the marbles.

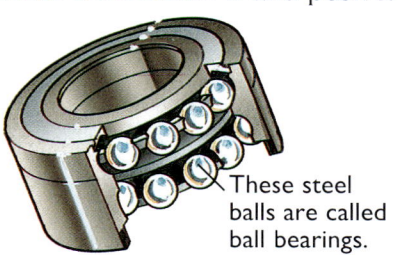

These steel balls are called ball bearings.

Wind tunnel

One of the most important tools used in aerodynamics is a wind tunnel. It allows scientists to measure and record exactly how air travels around a car at different speeds. A modern wind tunnel can also mimic extreme weather to see how parts of the car react. For example, strong jets of water can mimic very heavy rain to see if the car has any tiny leaks.

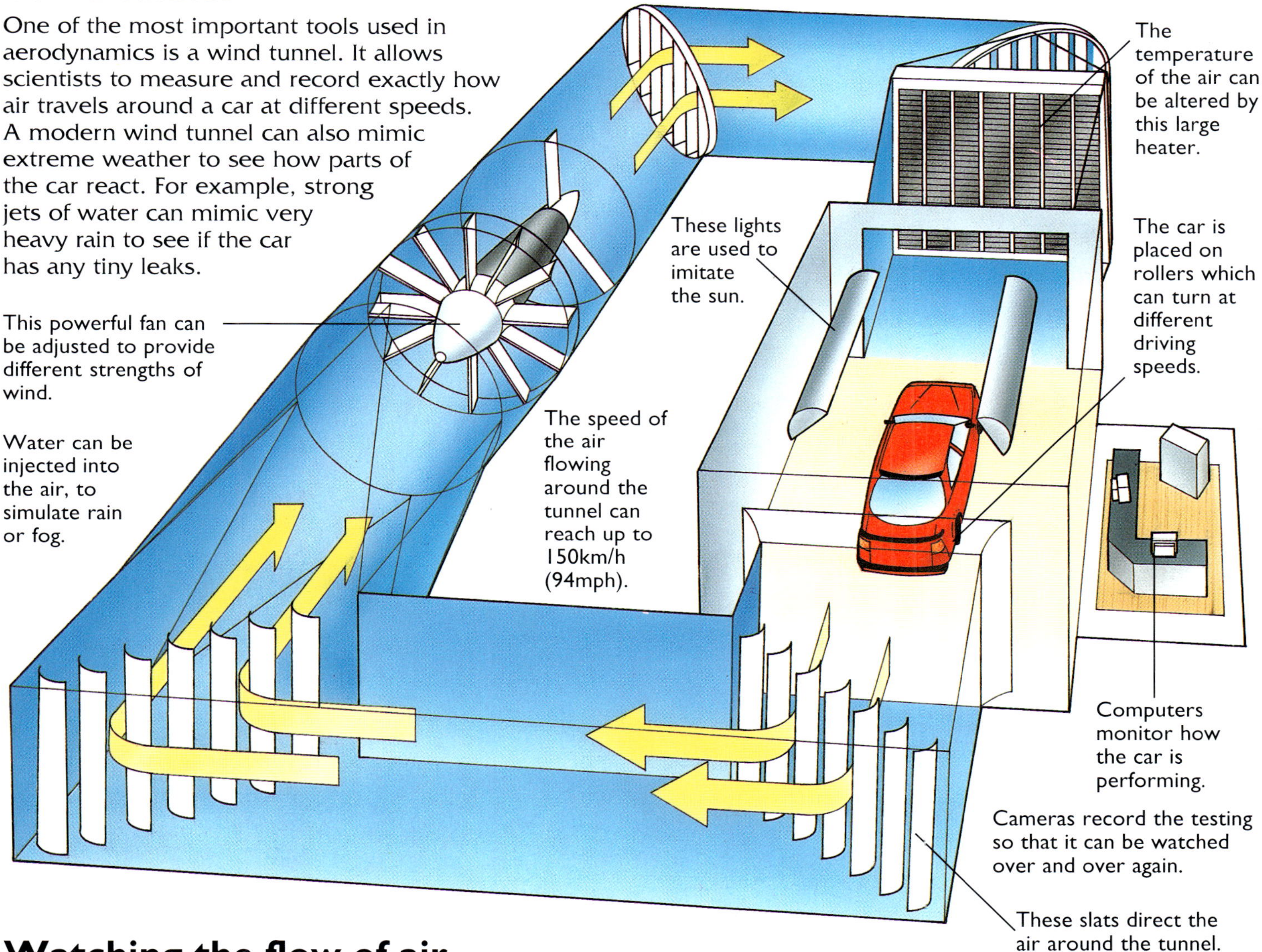

This powerful fan can be adjusted to provide different strengths of wind.

Water can be injected into the air, to simulate rain or fog.

These lights are used to imitate the sun.

The speed of the air flowing around the tunnel can reach up to 150km/h (94mph).

The temperature of the air can be altered by this large heater.

The car is placed on rollers which can turn at different driving speeds.

Computers monitor how the car is performing.

Cameras record the testing so that it can be watched over and over again.

These slats direct the air around the tunnel.

Watching the flow of air

Air is invisible, but its journey over and around a car needs to be watched and recorded. So, engineers have had to invent special techniques to see the air.

This man is injecting white gas into the air just before it flows around the car. Engineers can then see how the air travels.

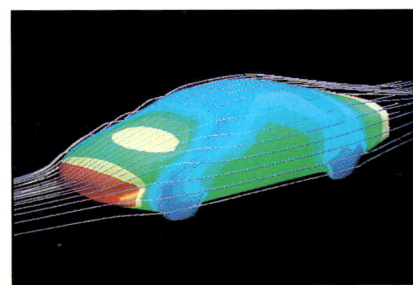

This computer is imitating the flow of air over a model car. Computers have greatly helped to improve cars' aerodynamics.

Vintage racing car

Almost as soon as cars were invented, they were raced against each other. The first official race was in France in 1895. In the early part of the twentieth century, cars built solely for racing first appeared. They competed in famous races such as the Mille Miglia in Italy and the 24 hour race in Le Mans in France. Many of the original cars have been restored and can be found in museums or even racing on tracks at vintage car rallies.

Bentley 4½ Litre

This famous vintage racing car was first built in 1930. Earlier versions of the car raced throughout the 1920s and won Le Mans every year between 1927 and 1930.

The name 4½ Litre refers to the *volume** of the engine's cylinders all added together. This measure is called cubic capacity and is measured in cubic centimetres or cc.

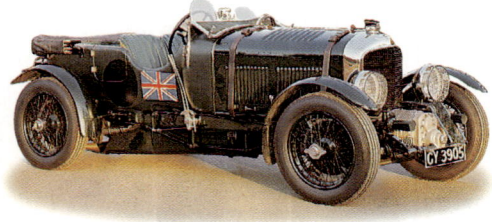

This is a Bentley 4½ Litre from the National Motor Museum in Beaulieu, England.

Le Mans today

The Le Mans race still takes place today but the cars racing in the competition have changed greatly since cars like the Bentley 4½ Litre raced there. For example, the first winning Bentley in 1927 raced at an average speed of 98km/h (61mph). The 1993 winner (shown right) raced at an average speed of 214km/h (132mph).

Modern light materials and a smaller but more powerful engine means that the Peugeot weighs just over a third of the weight of the Bentley so it goes faster. The Peugeot's top speed is over 400km/h (250mph). That is twice as fast as the Bentley.

The fuel tank is exposed which could be very dangerous if another car crashes into the back.

This hood can be lifted up to cover the driver and passengers in bad weather.

This mudguard is like the ones on bicycle wheels.

The driver and passenger used to slide around on these seats as the car had no seat belts.

This is a Peugeot 905B-EV11. It won the Le Mans race in 1993.

The SA35 engine in the Peugeot delivers over 600bhp (see page 11), that's almost three times the power of the Bentley engine.

This duct directs air onto the brakes to keep them cool.

The car's body comes off in sections to make it easy for the mechanics to get to the parts they want to repair.

This small piece of glass acts as a windshield for the passenger in the front seat. It is called an aeroscreen, because a similar screen was first used on aircraft.

The driver can flip up this wire mesh screen to protect himself from stones thrown up by cars in front.

Racing in the 1930s

Bentleys were just one of the famous makes of cars that raced in the 1930s. Others included Bugatti, Alfa Romeo and Mercedes-Benz. Racing cars of this time were usually open-topped. To keep warm, clean and dry, drivers wrapped themselves up in many layers of clothes and often a final covering of leather coats, hats, gloves and goggles.

This is a Bentley racing at Le Mans over 60 years ago.

The spare tyre is bolted on to the outside of the car body.

These slots in the car body are called vents. They let air inside the body to cool the engine down.

The leather strap stops the engine cover from flying open.

The Bentleys were originally painted in this distinctive and famous shade of green. It is called British racing green.

This is the Bentley symbol found on all Bentley cars.

The front headlight has a wire mesh cover to stop stones from breaking the glass.

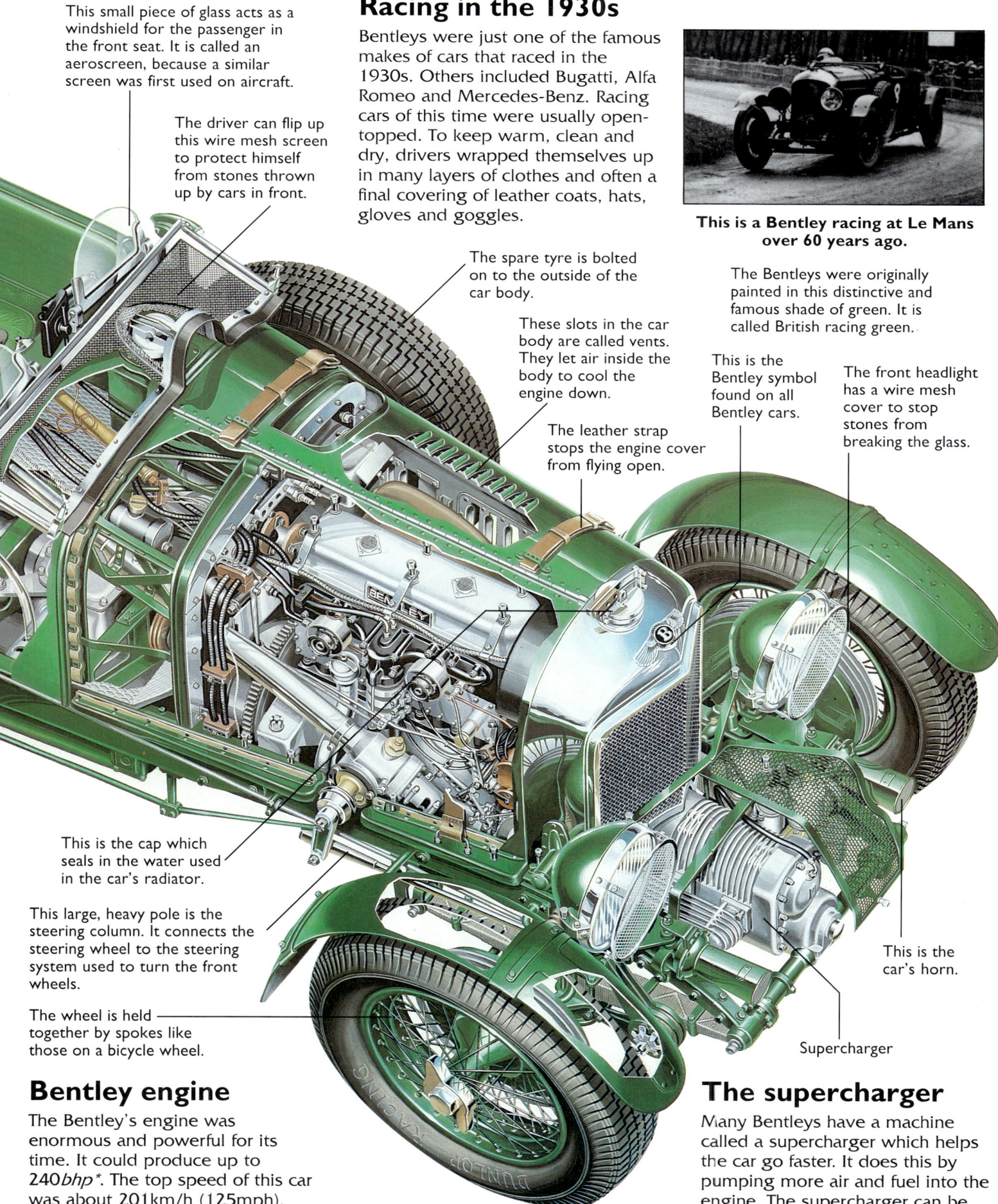

This is the cap which seals in the water used in the car's radiator.

This large, heavy pole is the steering column. It connects the steering wheel to the steering system used to turn the front wheels.

The wheel is held together by spokes like those on a bicycle wheel.

This is the car's horn.

Supercharger

Bentley engine

The Bentley's engine was enormous and powerful for its time. It could produce up to 240 *bhp**. The top speed of this car was about 201km/h (125mph).

The supercharger

Many Bentleys have a machine called a supercharger which helps the car go faster. It does this by pumping more air and fuel into the engine. The supercharger can be switched on and off by the driver.

Suspension

A car's suspension makes travelling much more comfortable. The car's wheels bounce up and down over holes and bumps in the road. The suspension system stops the whole car from bouncing around uncontrollably. The suspension also helps the wheels to stay in touch with the ground as much as possible. This improves the car's handling which is how drivers describe how easy a car is to control. There are several types of suspension. Many use a spring like the one shown on the right.

Spring

Damper (see below)

Active suspension

This modern system of suspension uses *hydraulic** cylinders instead of springs and dampers. The height of each wheel is controlled by the cylinders which are connected to a central computer in the car. Active suspension greatly improves a car's handling.

When accelerating, active suspension stops the front wheels from rising off the ground.

When braking sharply, active suspension stops the front of a car from dipping down.

When cornering sharply the inner wheels tend to rise. Active suspension helps keep them on the ground.

Riding on springs

When a car goes over a bump, the spring compresses, squashing up to absorb the energy of the bump; but it must eventually return to its normal position. A spring expands as quickly as it compresses. It expands past its normal size then is pulled back.

 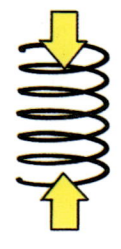  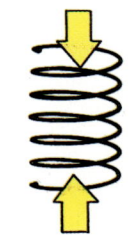

Normal size   Compressed   Expanded   Compressed

Springs rise and fall many times before they get back to normal size. A car body fitted only with springs would bounce up and down for quite some time. The solution is to fit each spring with a device called a damper.

Car body with just springs for suspension.

Car body rises and falls for a long time.

You can see the springs and dampers connected to each wheel on this picture.

Car suspension springs have to be strong to cushion the weight of a loaded car as it goes over a bump.

The suspension springs are joined to this car separately from the dampers.

This is the rear *axle**.

How a damper works

A damper slows the rise of the spring. The most common type of damper uses a piston joined to the spring and a cylinder of oil. This is called a hydraulic damper. It works by using special openings called *valves**. Oil travels through these valves faster when the piston moves down the cylinder than when it moves up. This means that the spring can compress quickly but expands slowly.

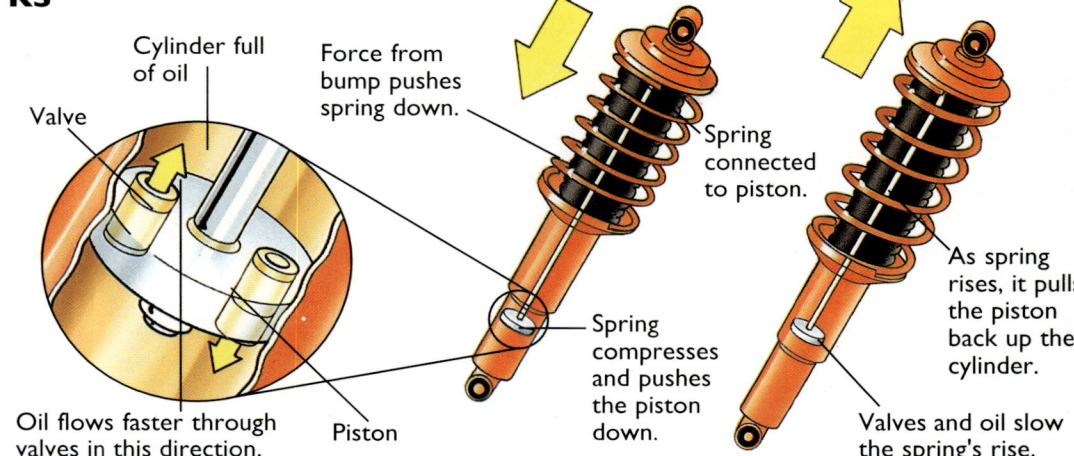

Cylinder full of oil

Valve

Oil flows faster through valves in this direction.

Force from bump pushes spring down.

Spring compresses and pushes the piston down.

Piston

Spring connected to piston.

As spring rises, it pulls the piston back up the cylinder.

Valves and oil slow the spring's rise.

20

Steering

Turning a car's steering wheel seems quite easy but your effort has to be increased so that you can turn the heavy car with its wheels gripping the ground strongly. Many cars use a system of two interlocking *gears** called a rack and pinion. At the end of the steering wheel column is the pinion. This interlocks with a sliding, toothed rail called the rack.

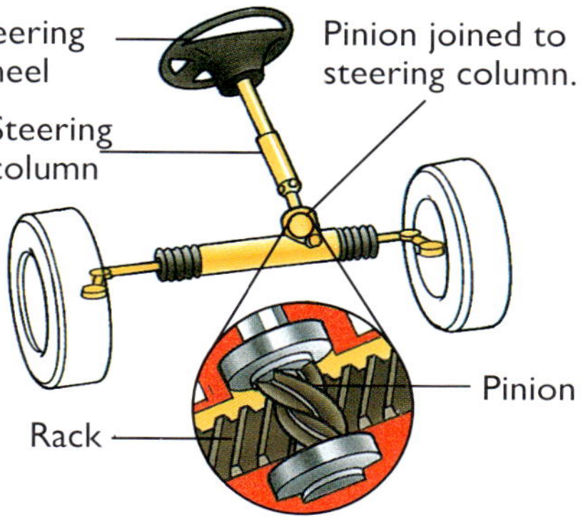

Steering wheel
Steering column
Pinion joined to steering column.
Rack
Pinion

The rack and pinion increase the turning force.

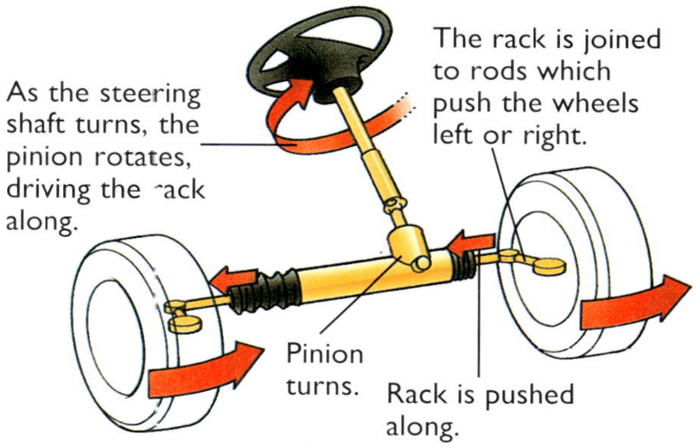

As the steering shaft turns, the pinion rotates, driving the rack along.

The rack is joined to rods which push the wheels left or right.

Pinion turns.
Rack is pushed along.

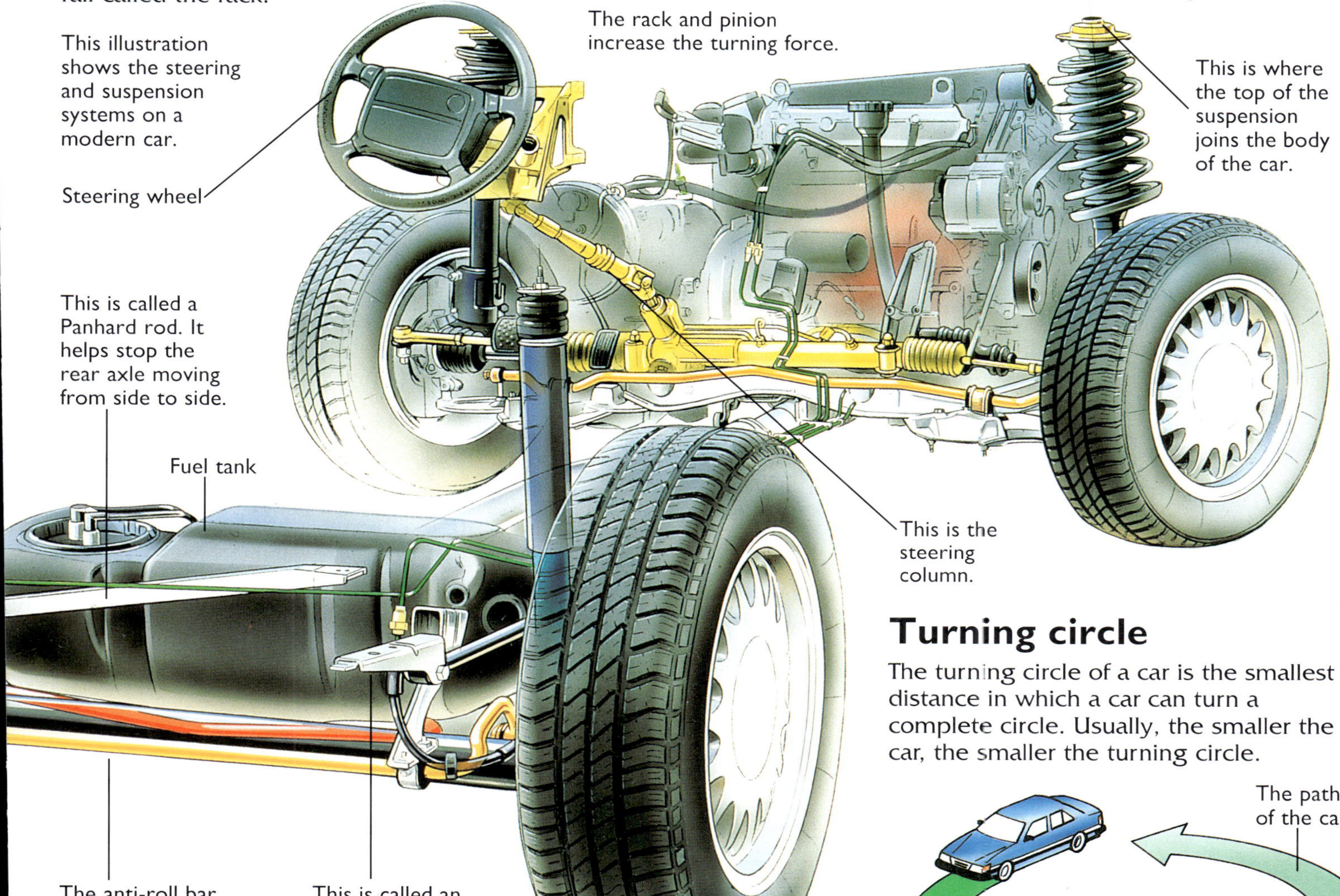

This illustration shows the steering and suspension systems on a modern car.

Steering wheel

This is called a Panhard rod. It helps stop the rear axle moving from side to side.

Fuel tank

This is where the top of the suspension joins the body of the car.

This is the steering column.

The anti-roll bar (see below) twists and helps keep the wheels on the ground.

This is called an arm. It joins the suspension to the car body.

Turning circle

The turning circle of a car is the smallest distance in which a car can turn a complete circle. Usually, the smaller the car, the smaller the turning circle.

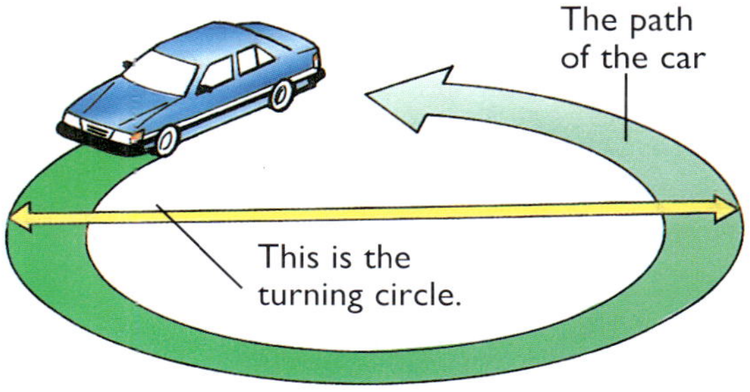

The path of the car
This is the turning circle.

Anti-roll bar

An anti-roll bar is a steel rod connected to the frame of the car. It helps prevent a car from leaning over on tight, fast corners or after hitting a large bump in the road. It does this by stopping the wheel from rising too high in the air. On the right you can see how it works by imitating its movement with a long, plastic ruler.

As one wheel on an axle goes up further than the other, the anti-roll bar starts to twist.

Twist ruler as shown.

Car wheel lifts up.

The bar acts like a spring, trying to keep each wheel at the same height.

Loosen grip and ruler untwists.

Wheel forced back down.

Brakes

Braking is all about stopping wheels that are spinning very fast. A modern car has a brake on each wheel and these all work at the same time when you press the brake lever. There are two main types of brake, disc and drum brakes. Both types use a *hydraulic** system. You can find out about hydraulic systems in planes on page 38.

Hydraulic system

Modern brakes are powered by a liquid put under pressure, called a *hydraulic** system. With the help of the car engine, the liquid is forced through pipes and cylinders to make the brakes move.

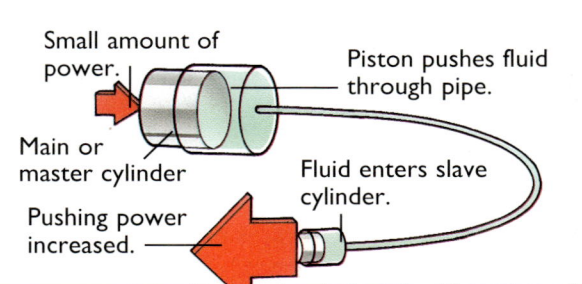

Disc brakes

Disc brakes are similar to brakes on a bicycle. Bicycle brakes have a lever called a caliper which opens and closes. On the ends of the caliper are brake pads which press hard onto the rim of the bicycle wheel. The *friction**, or rubbing, created by the pads slows the wheel down. When a car disc brake works, hydraulic fluid forces the caliper to close onto a disc joined to the car wheel.

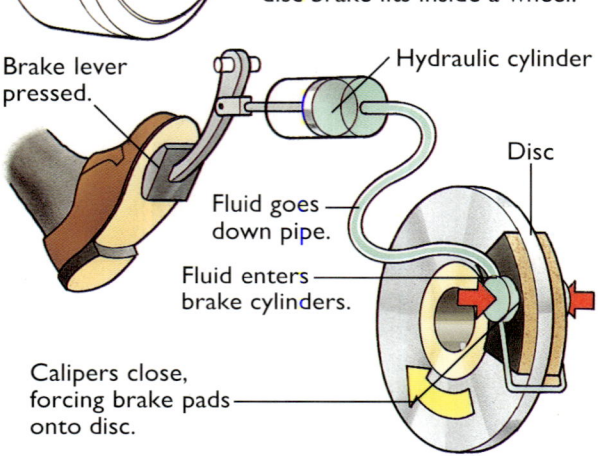

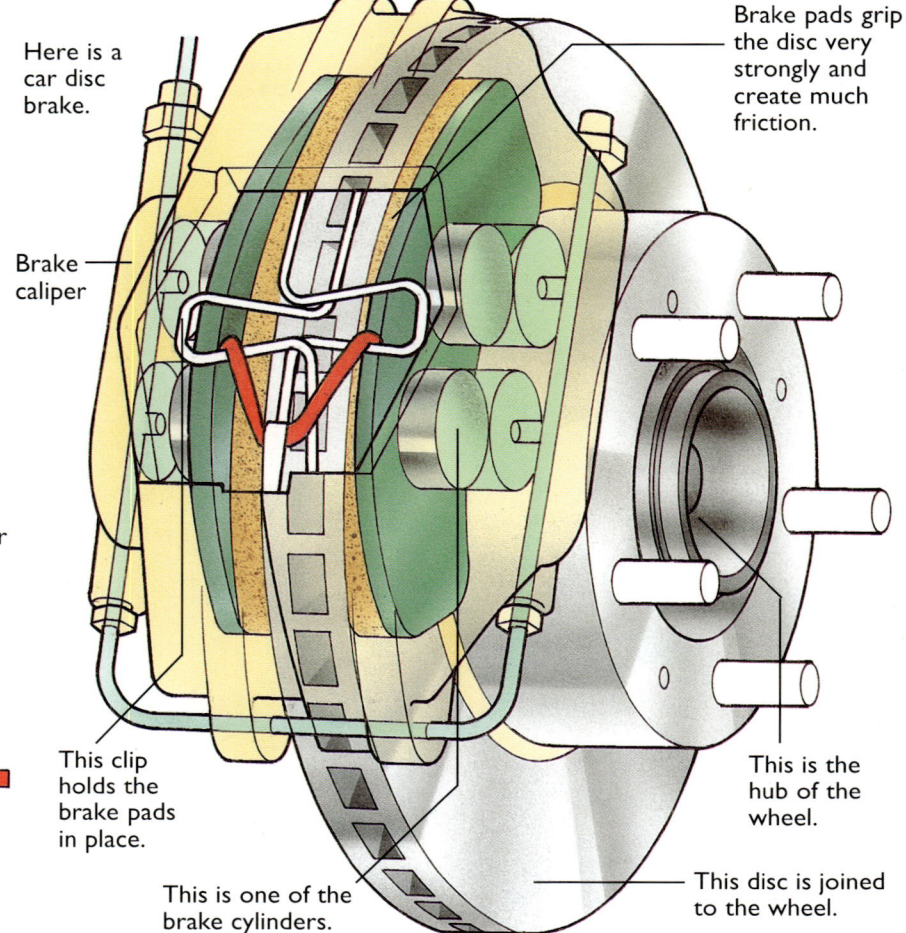

Drum brakes

Drum brakes are so-called because they have a metal drum attached to the car wheel instead of a disc.

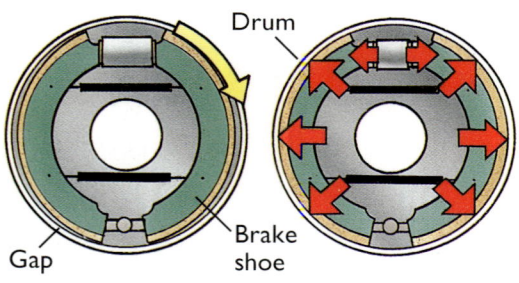

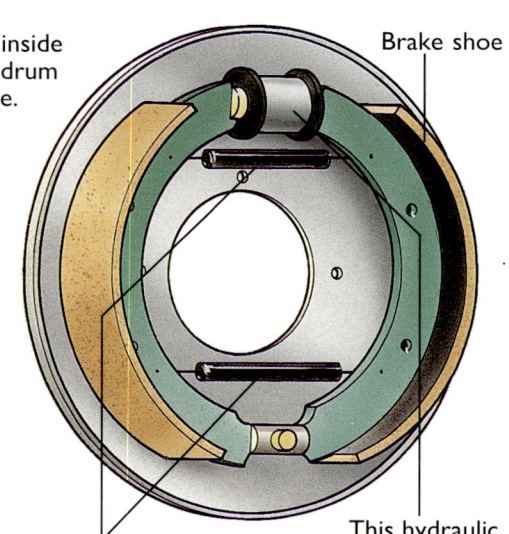

Brake pads fit on curved brake shoes which sit inside the drum. There is a gap between the shoes and drum.

When the driver brakes, a hydraulic system pushes the shoes out to touch the drum. This slows the wheel down.

Powerful springs push the shoes away from the drum when the brake is released.

This hydraulic cylinder pushes the brake shoe onto the drum.

ABS braking

Sometimes, if brakes are pressed too hard, they can lock and stop the wheels from spinning. If the wheels are not going around, the driver cannot steer and may skid and hit something. ABS, or anti-locking brake system, has sensors on the wheels linked to a computer. This can adjust the brake pressure up to 10 or 12 times a second. This means the brakes are put on and off a fraction all the time; enough to allow the driver to brake and steer safely around hazards.

Tyres

Modern tyres are complicated car parts, designed by computers and tested as thoroughly as, and sometimes more than, other parts of the car. So much work goes into designing and making them because they are vital to a car's safety and performance. Tyres are the only part of the car that touch the ground. They must be able to grip the road enough to move the car along efficiently and allow the driver to control the car easily both in a straight line or when turning.

This is a modern car tyre, made by Michelin. It is made up of many different layers called plies.

Types of tyre

There are many different types of tyres for different vehicles and driving conditions.

The first tyres were made of solid rubber. This tyre from a Bugatti car of the 1930s is filled with air like a modern one but is not as wide.

This tyre is made for rallying across heavy mud or sand. It has a very deep, chunky tread, a lot like tyres used on tractors.

This tyre is very wide and has almost no tread. It is used by Grand Prix cars when the track is completely dry. If it rains, the tyre must be swapped for one with more tread.

Radial tyres*, like this one, have plies running at right angles to the side walls.

The top surface of a tyre is marked with a pattern of grooves called the tread.

The tread clears water, mud, dirt and snow out from between the tyre and the road.

This layer is called the undertread. It is made of tough rubber.

Tyre rubber is a mixture of natural rubber which can stand a lot of heat and man-made rubber which is more hard wearing.

This part of the tyre is made from thin steel wires all woven together. It makes the tyre stronger.

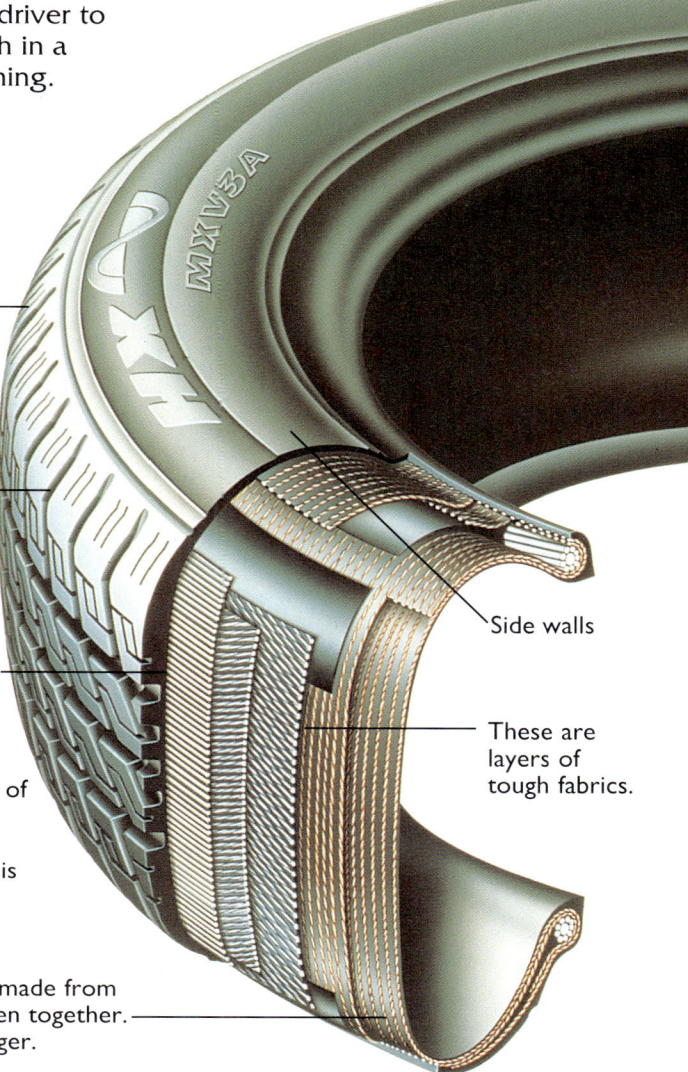

Side walls

These are layers of tough fabrics.

Grip in wet weather

If the grooves in the tread of a tyre are too narrow or too shallow, they cannot do their job properly. Instead of spraying water out from under the tyre so it can grip the road, a slippery film of water forms between the tyre and the road. The car can lose its grip and go out of control. This is called aquaplaning. Modern tread is designed to avoid this.

This new Goodyear tyre has a large channel down its middle to remove more water than an ordinary road tyre.

Complicated tread pattern

This special photo shows water being directed out from under the tyre.

The part of the tyre you can see above is the footprint. That is the part which touches the ground at any one time.

Getting rid of tyres

Millions of worn down tyres go for scrap every year. Because each tyre is made from a range of different materials, they are hard to recycle. Some tyres are burned but many are buried underground where they can create pollution or help fuel dangerous underground fires.

There are other, less harmful ways to dispose of scrap tyres. For example, some can be given a new layer of tread in a complicated process called retreading or remoulding.

These parts of the car bumper are made from scrap tyres.

Sports car

Sports cars are the quickest cars on the road. They can start off very fast and steer precisely and accurately. They are great fun to drive, but are often small inside and extremely expensive to buy. Some drivers race their sports cars on tracks. Famous sports car makers include Ferrari, Lotus, Chevrolet and Jaguar.

The 911 has been changed many times. However, it still looks similar to this early model.

Porsche 911 Carrera 4

First built in 1963, the Porsche 911 is one of the best-known sports cars. It has raced at Le Mans and won the Monte Carlo rally. Over 250,000 have been built. This Carrera 4 model is one of the latest of many versions of the 911.

Here's the spare wheel.

This fan cools the oil used in the engine.

Rounded design

Dr Ferdinand Porsche designed one of the world's most popular cars many years before the Porsche company built the 911. The two cars have a rounded shape. Can you guess what the other car is? Turn to page 32 for the answer.

This car has its fuel tank at the front.

This drive shaft takes power from the engine to the car's wheels.

Four wheel drive

In most cars the engine feeds power either to the back or front pair of wheels. The wheels that do not receive power are pushed or pulled by the others. In a four wheel drive car, all four wheels receive power straight from the engine. This helps the car grip the road better and makes it easier for the driver to control it in bad weather.

The engine's power can be varied between the back and front wheels, depending on which can grip the road better. In normal driving, the back wheels get 70% and the front wheels, 30%.

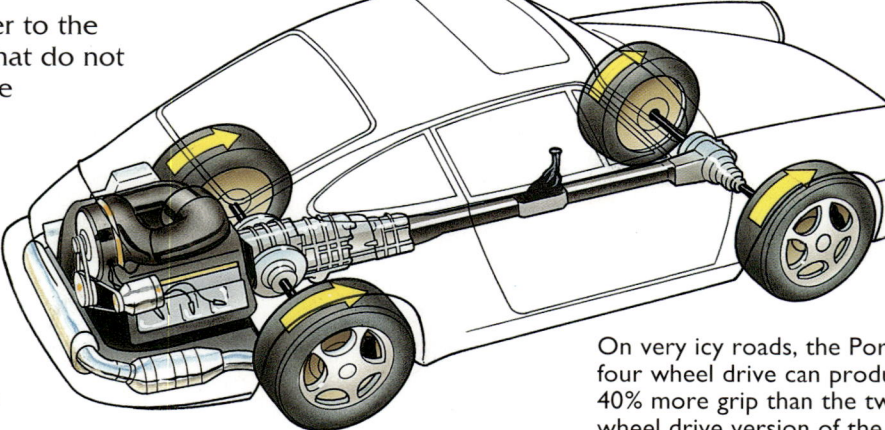

On very icy roads, the Porsche four wheel drive can produce 40% more grip than the two wheel drive version of the car.

Acceleration

Acceleration is how fast a car can increase its speed. Sports cars must have good acceleration. This is often measured as how quickly a car can go from standing still to 100km/h (62mph). The 911 Carrera 4 can do this in 5.7 seconds.

This Porsche 959 can accelerate from 0 to 100km/h (62mph) in just 3.9 seconds.

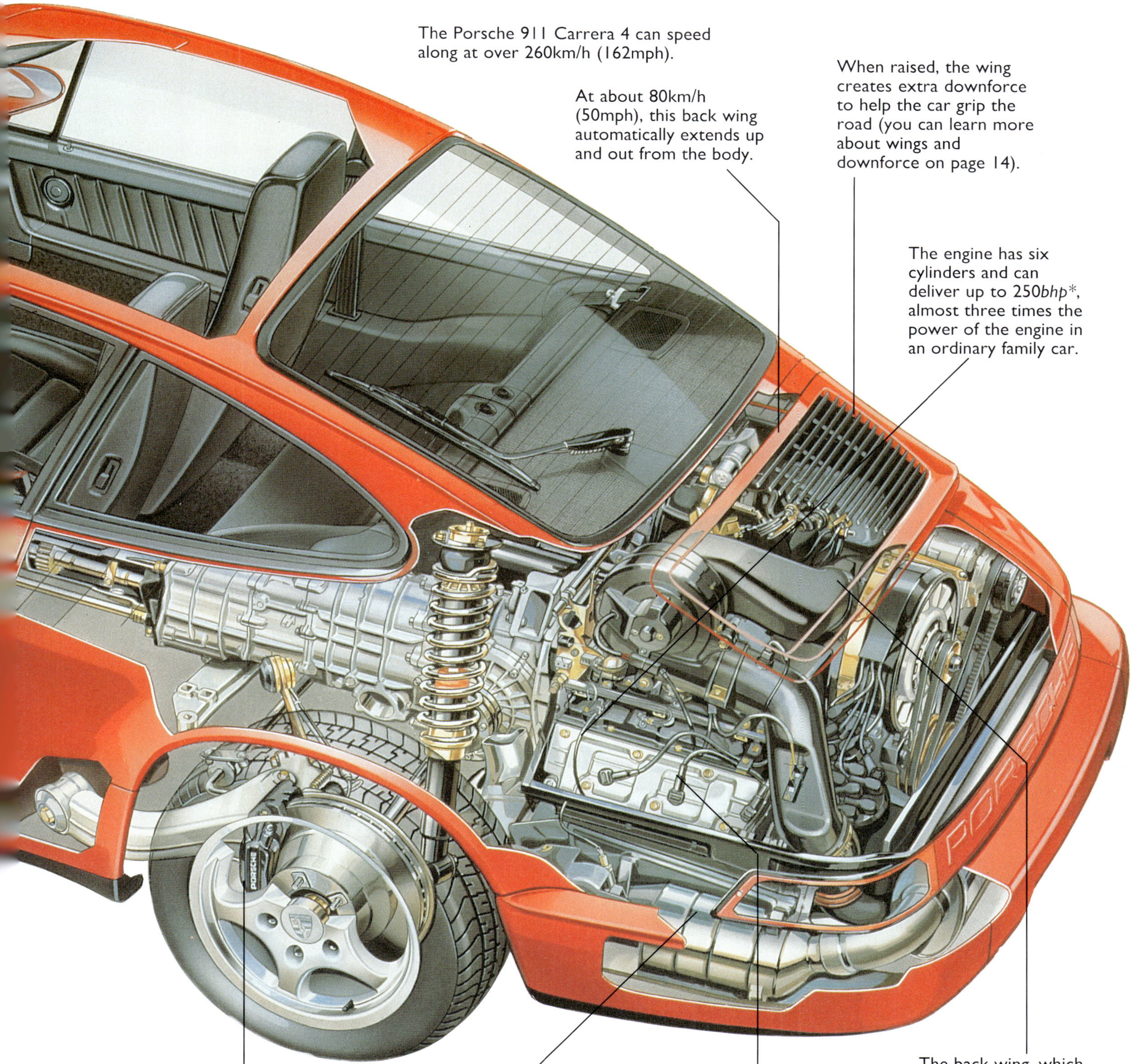

The Porsche 911 Carrera 4 can speed along at over 260km/h (162mph).

At about 80km/h (50mph), this back wing automatically extends up and out from the body.

When raised, the wing creates extra downforce to help the car grip the road (you can learn more about wings and downforce on page 14).

The engine has six cylinders and can deliver up to 250*bhp**, almost three times the power of the engine in an ordinary family car.

Unlike many family cars which have drum brakes on the back wheels, the Porsche has anti-locking disc brakes on all four wheels.

This is the Porsche's catalytic converter. It helps cut down pollution from the engine.

Each cylinder has not one, but two spark plugs. This is called dual *ignition** and helps the engine run more smoothly.

The back wing, which is lifted up at faster speeds, allows more air around the engine to cool it.

25

Electrical system

A modern car would not work without its electrical system. Electricity is needed to start the car, to make the spark plugs burn the air and fuel mixture in the engine and to power the lights, windshield wipers and other electrical parts of the car.

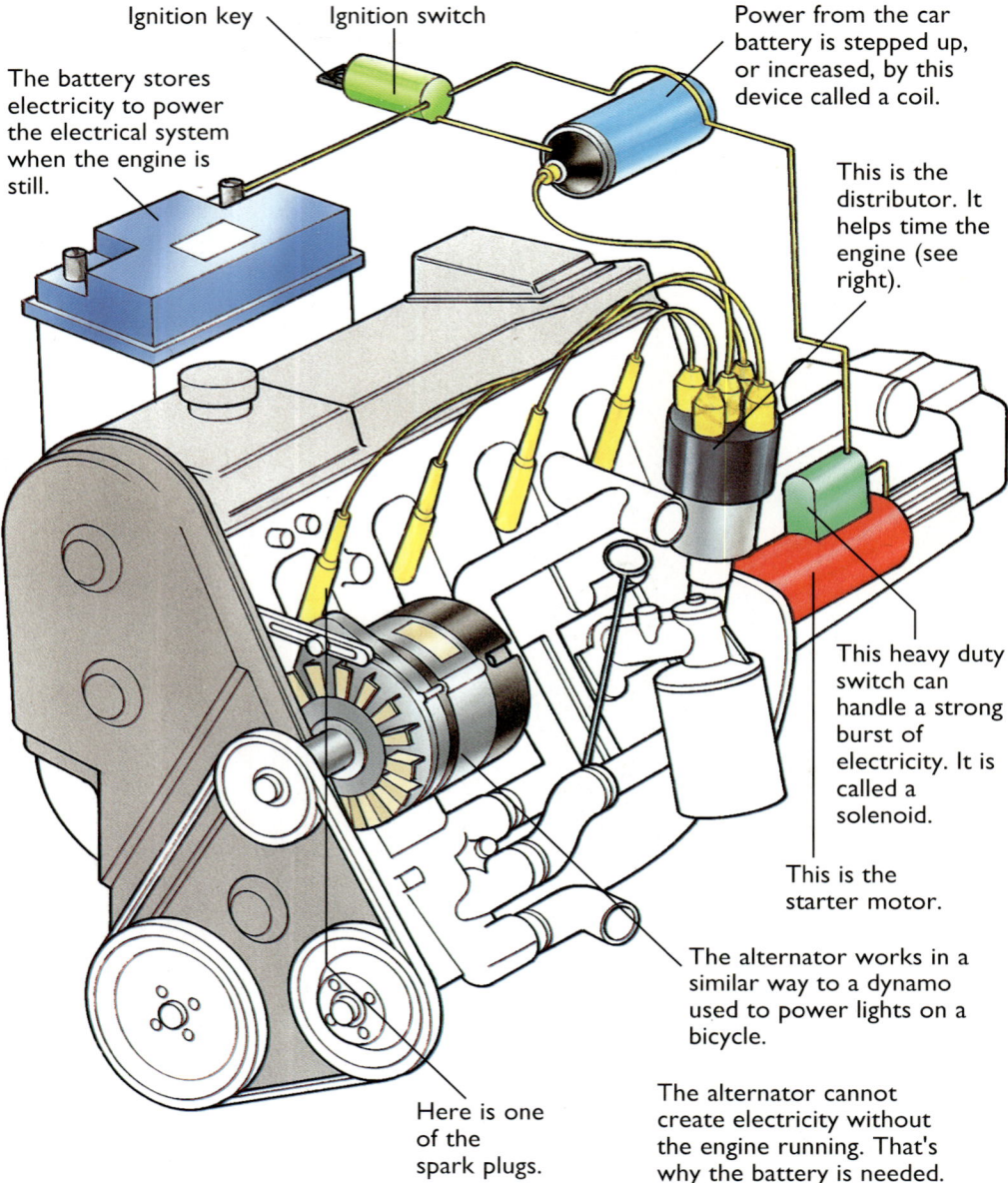

The battery stores electricity to power the electrical system when the engine is still.

Ignition key

Ignition switch

Power from the car battery is stepped up, or increased, by this device called a coil.

This is the distributor. It helps time the engine (see right).

This heavy duty switch can handle a strong burst of electricity. It is called a solenoid.

This is the starter motor.

The alternator works in a similar way to a dynamo used to power lights on a bicycle.

The alternator cannot create electricity without the engine running. That's why the battery is needed.

Here is one of the spark plugs.

Starting the engine

Turning the ignition key releases electricity from the battery, which passes through the solenoid switch to the starter motor.

The starter motor turns the engine's flywheel making the pistons move up and down the cylinders. At the same time, more electricity is passed to the coil.

The coil increases the electricity's voltage. The electricity is then fed to the distributor which times the electricity reaching the spark plugs.

As the engine runs, it turns the alternator. The alternator uses this movement to create electricity, which it feeds to the battery and uses to run the electrical system.

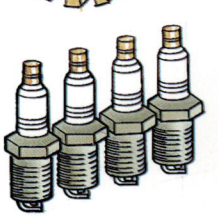

The spark plugs start producing sparks which set light to the air and fuel mixture in the cylinder.

Lights

Lights are needed for driving in fog, bad weather conditions and at night. All modern cars have many different lights but they were considered luxury accessories until the 1930s. The first lights were powered by gas like old street lamps but today, all car lights are powered by electricity.

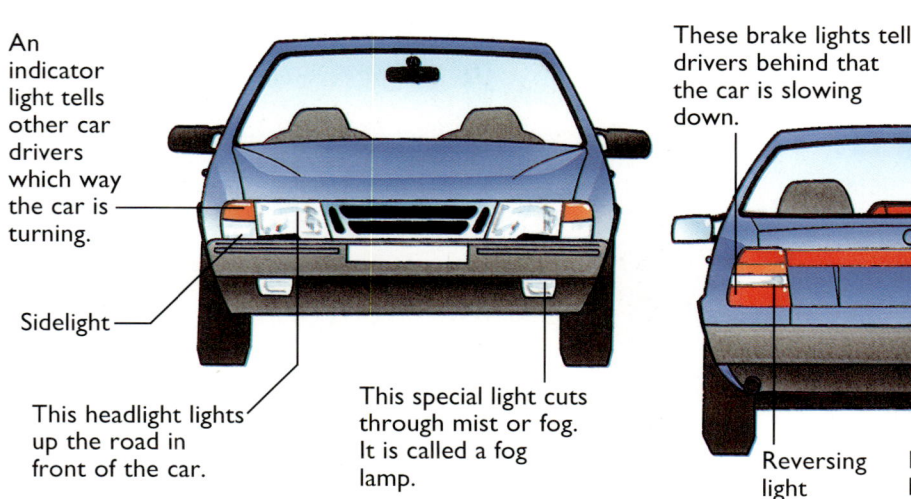

An indicator light tells other car drivers which way the car is turning.

Sidelight

This headlight lights up the road in front of the car.

This special light cuts through mist or fog. It is called a fog lamp.

These brake lights tell drivers behind that the car is slowing down.

Some new cars have extra brake lights in the rear window.

Reversing light

Rear indicator lights

26

Timing the engine

The power from the coil has to reach the spark plugs at precisely the right time. If it doesn't, then the engine will run very poorly. Getting the timing right is the job of the distributor.

At the middle of the distributor is an arm which turns at a speed decided by how fast the engine is working. Each time the arm gets close to a metal point, it completes an electrical circuit which passes electricity on to the spark plug.

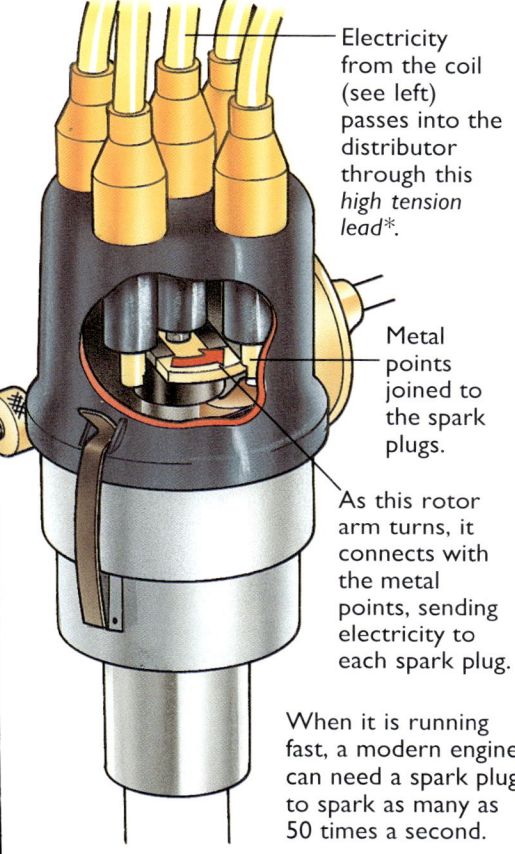

Electricity from the coil (see left) passes into the distributor through this *high tension lead**.

Metal points joined to the spark plugs.

As this rotor arm turns, it connects with the metal points, sending electricity to each spark plug.

When it is running fast, a modern engine can need a spark plug to spark as many as 50 times a second.

The car's instruments

The dashboard is the panel in front of the driver and it holds many of the instruments and controls used to drive a car. Some cars now have a screen which displays maps of where the car is heading. The screen is joined to a navigation computer which can offer suggested routes between two places. Nearly all cars have heaters and air fans to keep the driver and passengers at a comfortable temperature.

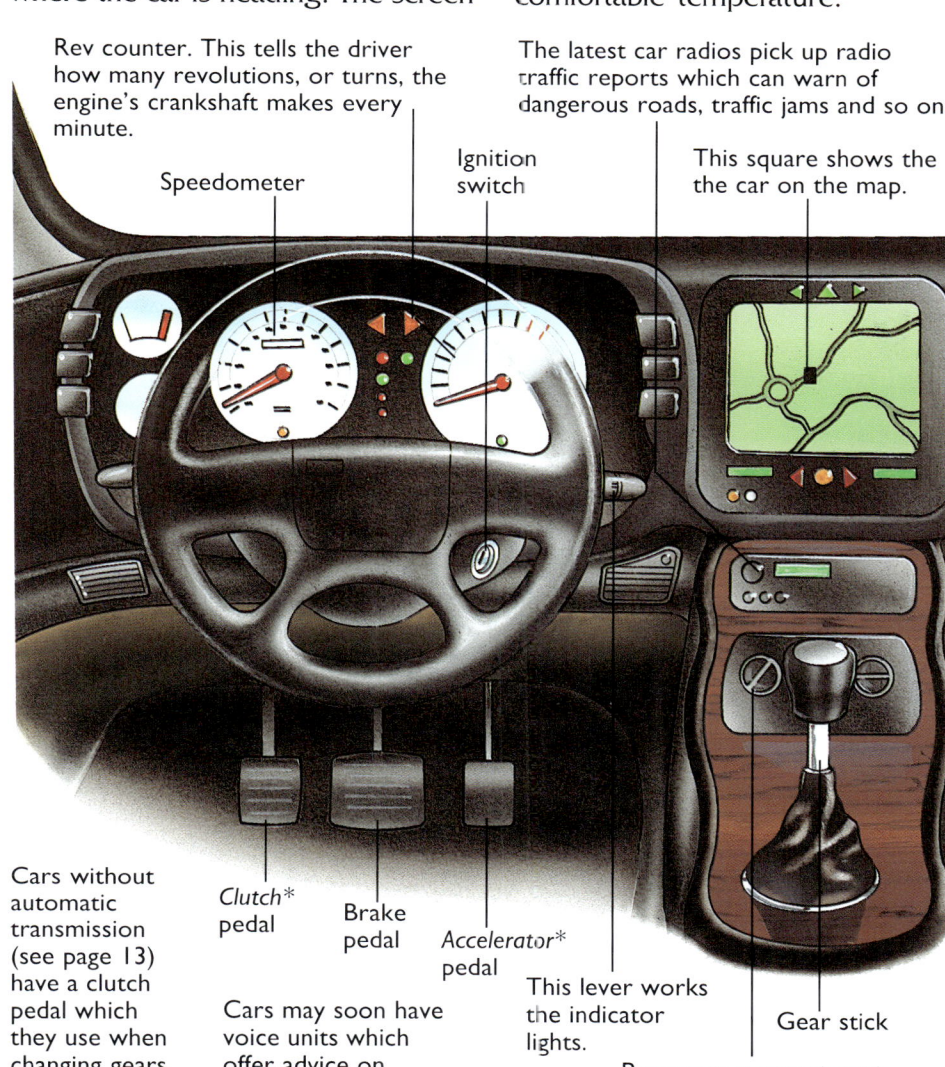

Rev counter. This tells the driver how many revolutions, or turns, the engine's crankshaft makes every minute.

The latest car radios pick up radio traffic reports which can warn of dangerous roads, traffic jams and so on.

Speedometer

Ignition switch

This square shows the the car on the map.

Cars without automatic transmission (see page 13) have a clutch pedal which they use when changing gears.

Clutch* pedal

Brake pedal

Accelerator* pedal

Cars may soon have voice units which offer advice on which way to go.

This lever works the indicator lights.

Gear stick

Passenger compartment heater controls.

Dipping headlights

Most countries have rules about the number, type and position of lights on a car. One of the most important rules is about dipping headlights. Full headlights are used when there are no other vehicles on the road. Bright headlights pointing straight ahead, however, can dazzle the driver of a car coming the other way. To prevent this, modern headlights are designed to dip their beam down onto the road at the flick of a switch.

Dipped lights

Full lights

On many cars, when the beam headlights are selected, the dipped beams stay on too.

This sports car has odd pop up lights. Do you know why? (see page 32 for the answer).

27

Off-road car

Cars are sometimes needed off roads to drive over rough ground. Off-road cars are as happy on the road as they are over grassland or on a bumpy dirt track.

Range Rover LSE

The Range Rover is a very popular off-road vehicle made by the Land Rover company. The LSE version shown here has a top speed of 180km/h (112mph) and weighs 2150kg (4740lb) when it is empty.

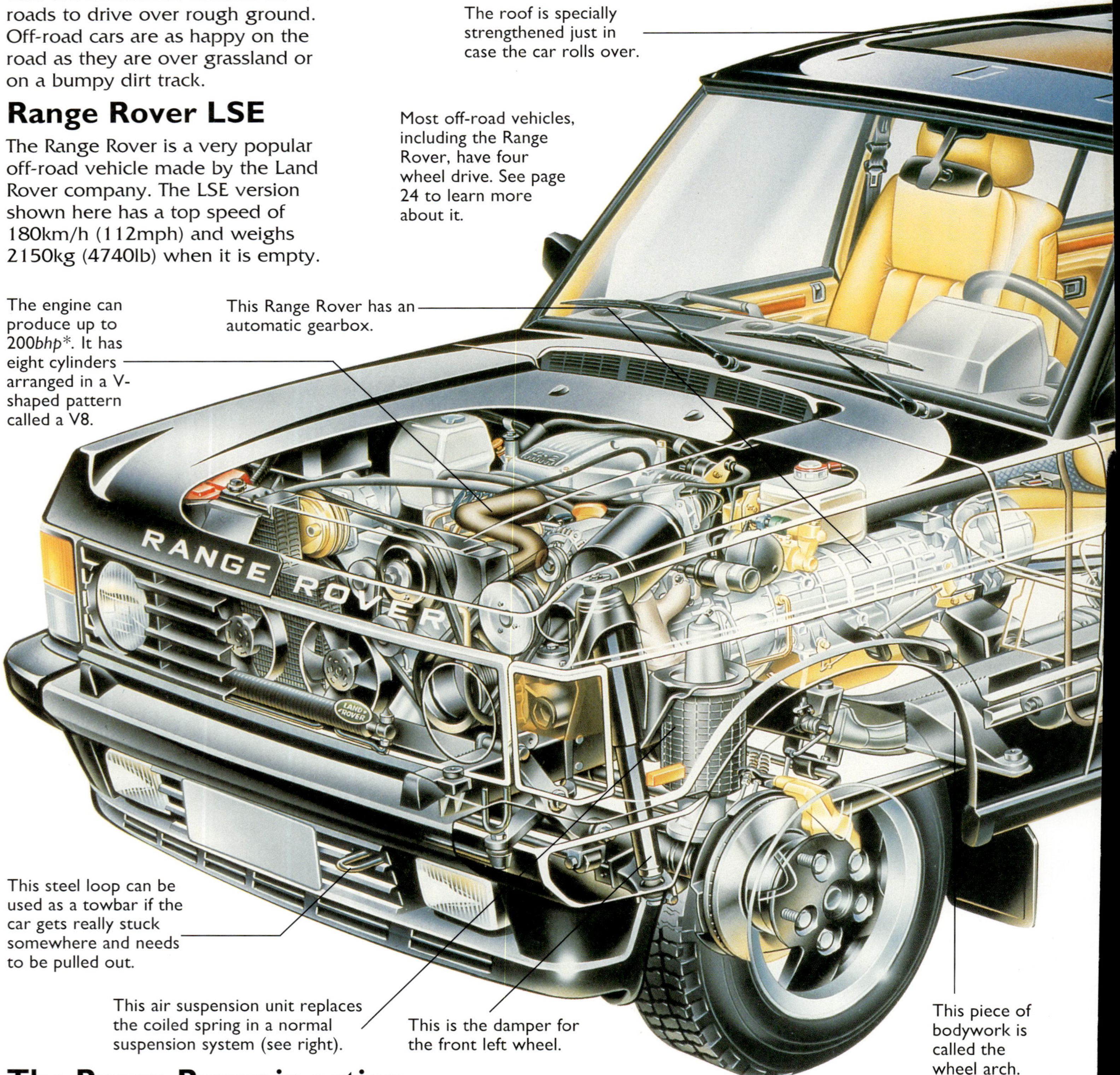

The roof is specially strengthened just in case the car rolls over.
Most off-road vehicles, including the Range Rover, have four wheel drive. See page 24 to learn more about it.
The engine can produce up to 200*bhp**. It has eight cylinders arranged in a V-shaped pattern called a V8.
This Range Rover has an automatic gearbox.
RANGE ROVER
This steel loop can be used as a towbar if the car gets really stuck somewhere and needs to be pulled out.
This air suspension unit replaces the coiled spring in a normal suspension system (see right).
This is the damper for the front left wheel.
This piece of bodywork is called the wheel arch.

The Range Rover in action

POLICE
K383MTU

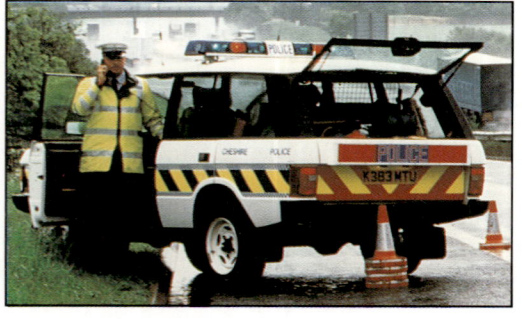

The police use the car for traffic control.

Off-road vehicles like the Range Rover are used for all sorts of jobs. Many police forces in Great Britain and abroad have Range Rovers. It is used as an ambulance and fire-fighting vehicle in some isolated areas. Farmers and forestry workers use the Range Rover for travelling through deep, muddy streams and over very rough ground.

RANGE ROVER

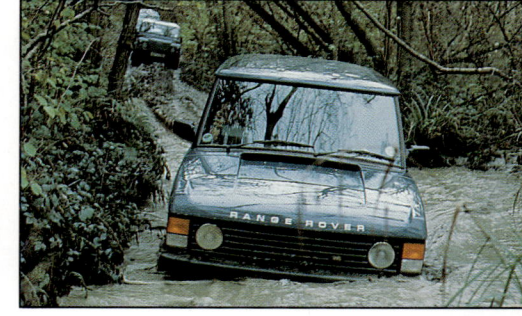

Range Rovers can cross deep streams.

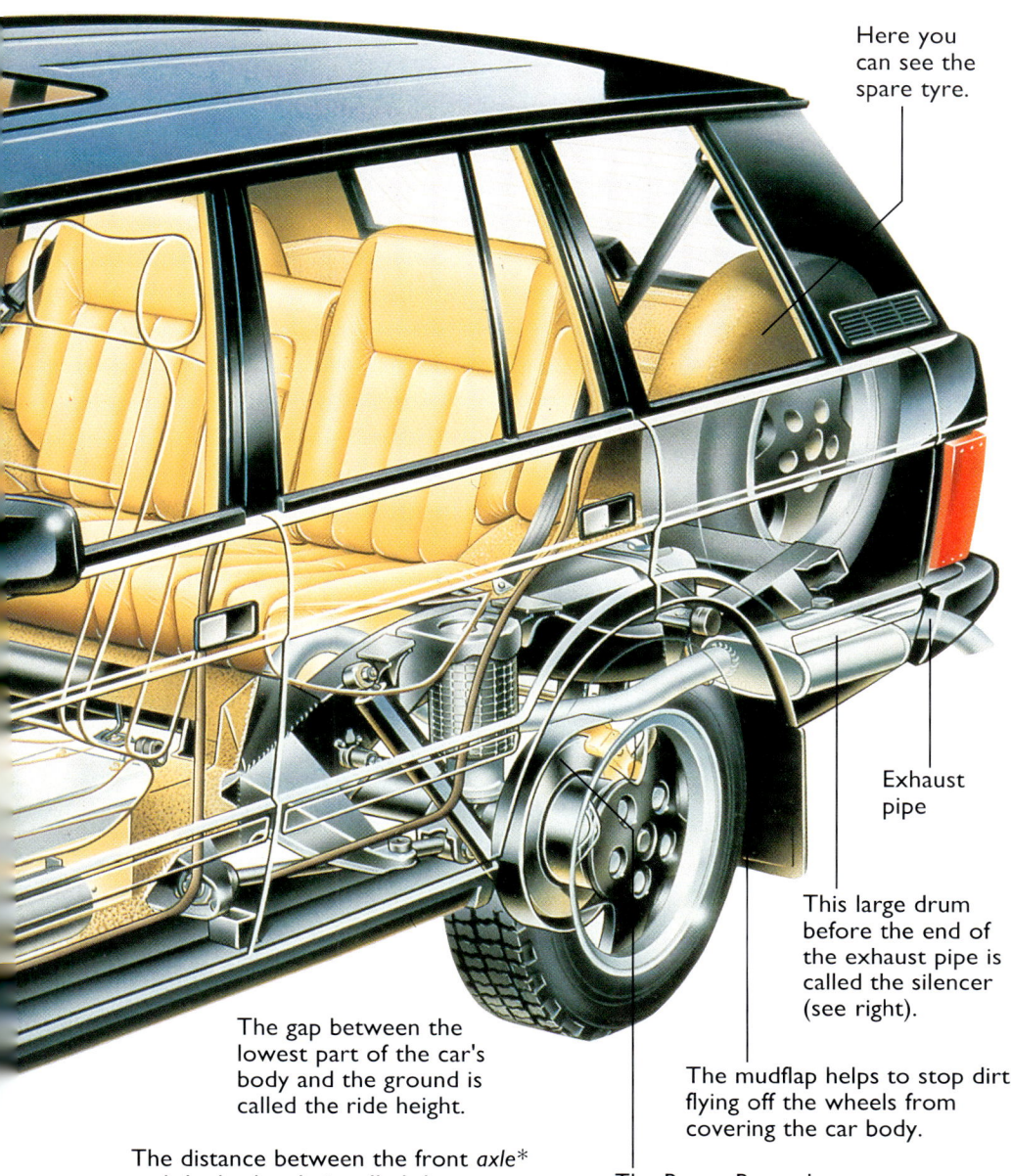

Here you can see the spare tyre.

Exhaust pipe

This large drum before the end of the exhaust pipe is called the silencer (see right).

The mudflap helps to stop dirt flying off the wheels from covering the car body.

The Range Rover has ABS anti-lock disc brakes (see page 22).

The gap between the lowest part of the car's body and the ground is called the ride height.

The distance between the front *axle** and the back axle is called the wheelbase. The wheelbase of this Range Rover is just over 2.7m (108in).

Air suspension

The Range Rover's suspension system uses powerful dampers and anti-roll bars, but it doesn't use normal springs (see page 20). Instead, *pneumatic** cylinders, full of pressurized air, are used to cushion the car body and absorb much of the energy from a bump or rut.

The suspension units on each wheel are independent. This means that each wheel can adjust its positioning as the car travels over uneven surfaces. Electronic sensors linked to a computer inside the car can adjust the gap between the car body and the wheels.

To clear obstructions, the body can rise up off the wheels.

For loading and unloading, the car body drops low over the wheels.

When going fast, the car lowers to improve its aerodynamics (see pages 16-17).

The exhaust system

The waste gases from the engine are piped out of the car by the exhaust system. As they leave the engine they are very hot and at a high pressure which would make a lot of noise if they went straight into the air. The exhaust system forces the gases on a long journey through pipes and a special device called a silencer (or muffler). By the time the gases leave the exhaust system they are much cooler and quieter.

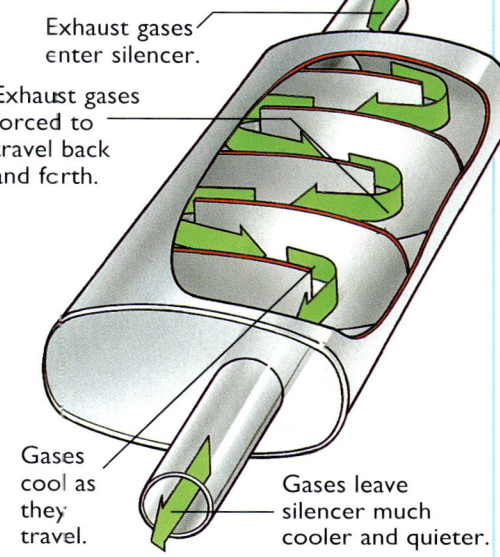

Exhaust gases enter silencer.

Exhaust gases forced to travel back and forth.

Gases cool as they travel.

Gases leave silencer much cooler and quieter.

Catalytic converter

The waste gases from the exhaust contain many substances which pollute the air. Many cars now use lead-free petrol as lead is a serious pollutant. More and more cars are also being fitted with a catalytic converter. This changes some of the waste gases that pass through it into other, less harmful chemicals. For example, dangerous nitrogen oxides are converted into harmless nitrogen and water.

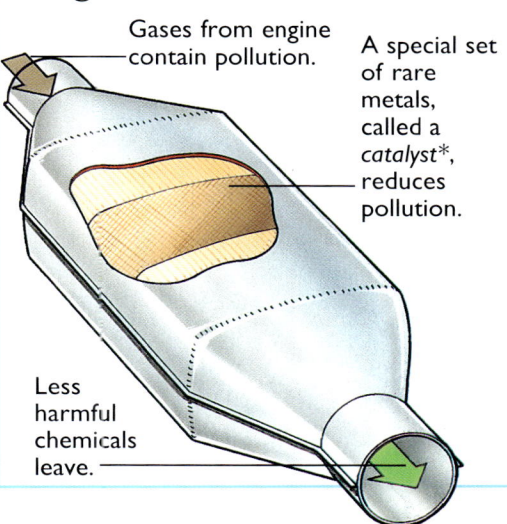

Gases from engine contain pollution.

A special set of rare metals, called a *catalyst**, reduces pollution.

Less harmful chemicals leave.

29

The future

Cars in the future will still need brakes, steering, suspension and all the other systems shown in this book. With the exception of a few top sports cars, future cars will not go much faster. But they are expected to be safer, easier to drive and less harmful to the environment. Here is a possible car of tomorrow with some of the futuristic features that it may include.

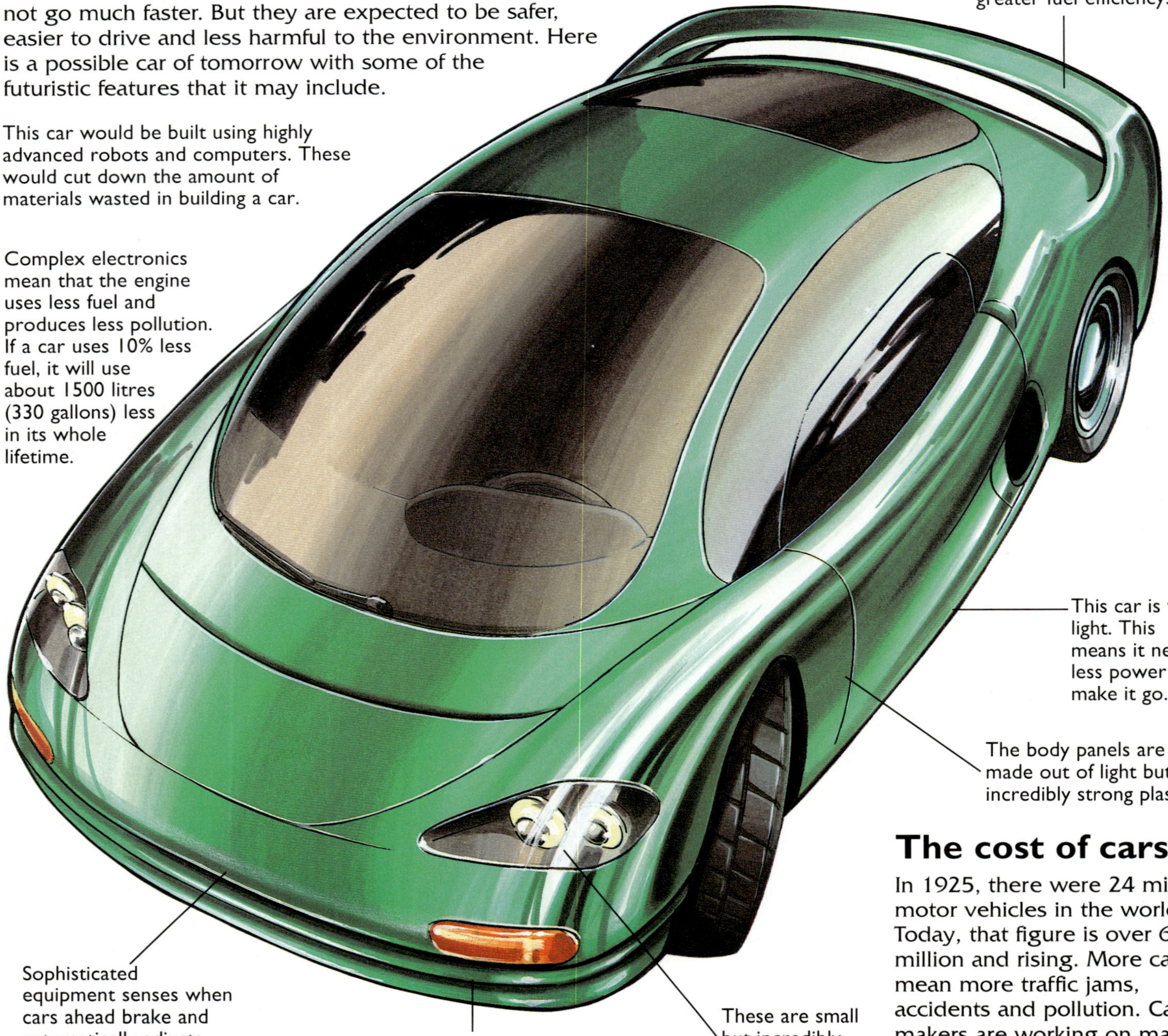

This car would be built using highly advanced robots and computers. These would cut down the amount of materials wasted in building a car.

Complex electronics mean that the engine uses less fuel and produces less pollution. If a car uses 10% less fuel, it will use about 1500 litres (330 gallons) less in its whole lifetime.

The car is very streamlined. This gives it better road handling and greater fuel efficiency.

This car is very light. This means it needs less power to make it go.

The body panels are made out of light but incredibly strong plastics.

Sophisticated equipment senses when cars ahead brake and automatically adjusts the car's speed.

These are small but incredibly powerful lights.

The cost of cars

In 1925, there were 24 million motor vehicles in the world. Today, that figure is over 600 million and rising. More cars mean more traffic jams, accidents and pollution. Car makers are working on many ways to reduce these things.

Safety

Safety researchers will find new ways to protect passengers in a crash. In fact, experts expect to find many new ways to prevent accidents in the first place.
Sensors will be able to scan the road ahead much better than a driver's eyes, especially in the dark and fog. Information from the cameras will appear on instruments called Head Up Displays (HUDs). The driver can see these without bending his head to look. At the moment, HUDs are only found in modern fighter jets. You can find out about these on page 57.

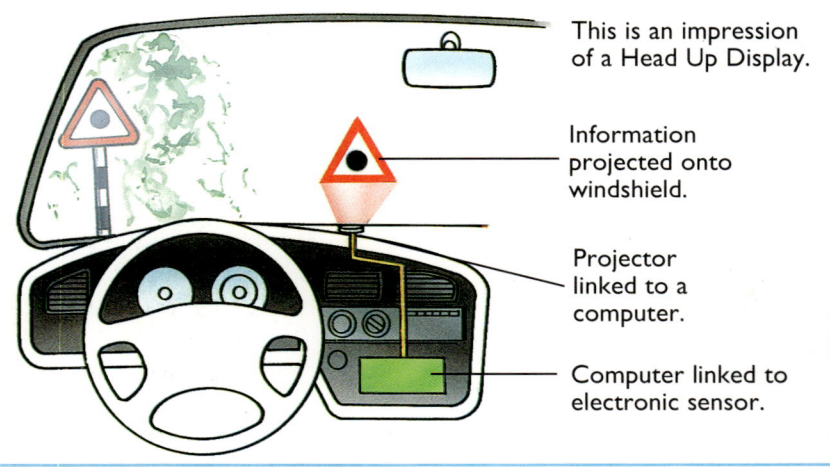

This is an impression of a Head Up Display.

Information projected onto windshield.

Projector linked to a computer.

Computer linked to electronic sensor.

Design your own

Virtual reality machines, which can show you exactly how it would look and feel inside your car, may allow you to design the interior to your own taste before the car is built.

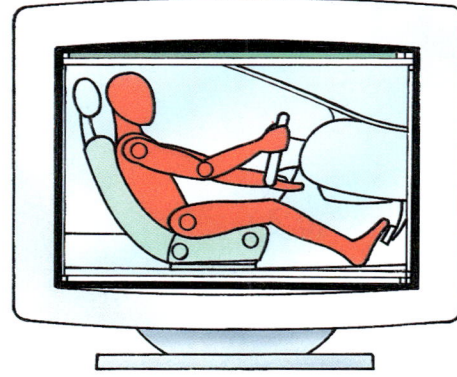

This dummy on the computer screen is testing the driver's seat. The dummy can be altered to the size of each car buyer.

Electric power

Electric cars, such as the one on page 11, are likely to become more common. Many will use a powerful battery which can be recharged, but some may be solar-powered like this three-wheeled car. It was built in Japan and called the Honda Dream.

Solar panels

It averaged a speed of 85km/h (53mph) when it won an Australian race for solar cars.

City driving

Special small cars for crowded towns and cities will become popular. Below is a city car being built by Ford. In the years ahead, similar-shaped cars may be powered by a small electric motor, have solar panels on their roof and be able to park themselves automatically without any help from the driver.

This is Ford's prototype of a city car, the Ka.

Glossary

Accelerator. The pedal that controls the speed of the engine and, so, the speed of the car.

Aerodynamics. The science of how gases, such as air, move over an object. Aerodynamics affects the way cars are designed and built.

Air pressure. The force with which air pushes against an object. Air pressure is increased by pushing air into a small space. This is called compressing.

Alternator. A device which generates electricity from the engine's movement.

Axle. The bar or rod on which the car's wheels turn.

Brake horsepower (bhp). One common measure of the power produced by an engine. It is used by many car manufacturers in their advertisements.

Cam. A small, usually oval, wheel which helps to convert turning movement into up and down movement.

Catalyst. A substance which changes how other chemicals react. A catalyst in a car's exhaust system is used to lower pollution.

Chassis. The framework of the car, usually made of steel. Most of the car's main parts, such as the engine and body panels, are attached to it.

Clutch. Operated by a pedal, it disconnects the gearbox from the engine long enough for the gear to be changed smoothly.

Crank. Something that helps to convert up and down movement into turning movement.

Cross ply tyres. Tyres with their plies running diagonally across in overlapping layers.

Distributor. A part of the engine which makes the spark plugs produce a spark at precisely the right time.

Differential. A set of gears that allow the car's wheels to turn at different speeds when the vehicle goes around a corner.

Fan belt. A belt which is powered by the engine and helps power the alternator and the engine cooling fan.

Firing order. The order in which the spark plugs produce sparks.

Fossil fuel. Fuels formed by dead plants and animals squashed together over millions of years in the same way that fossils are created. Coal and oil are examples of fossil fuels.

Friction. The resistance found when one surface moves and rubs against another surface.

Gears. Devices with grooves or teeth on their edges that mesh, or link, with each other. When they are powered they can turn each other around. Many gears are toothed wheels like those that drive a bicycle chain.

Handbrake. A hand-operated lever usually connected to the back brakes. It holds a car still when parked.

High tension lead. A thick lead which carries high voltage electricity to and from certain parts of the engine, such as between the distributor and the spark plugs.

Hydraulics. Using a liquid to transmit power from one place to another. A car's brake system uses hydraulics. Early hydraulic machines used water but most today use oil or other liquids that do not freeze as easily as water.

Ignition. Setting light to and burning the fuel and air mixture in the engine's cylinders.

Lubricant. A slippery liquid, such as oil, used to cover surfaces that rub together. The lubricant helps reduce friction. The process of using a liquid in this way is called lubrication.

Pneumatic. Using a compressed gas, usually air, to fill a container or transmit power.

Radial tyres. Tyres which have their layers or plies running across the tyre at a right angle to the rim of the wheel.

Streamlining. To shape a car's body so that it can move through the air as smoothly as possible.

Tachometer. This is also known as the rev. counter. It shows the engine's speed in the form of how many times the crankshaft turns around every minute.

Torque. The turning force from an engine.

Tuning. Adjusting the car's engine so that it performs at its best.

Valve. A device that acts like a door, opening, closing and controlling the flow of a liquid or gas through a pipe or tube.

Volume. The measurement of how much space an object takes up.

Wheel spin. When the car's tyres cannot grip the road properly. This often happens when the road is slippery or icy or when the tyres are in very poor condition.

Cars: Facts

SOME MOTORING FIRSTS

1769 The earliest vehicle related to the motorcar was a **steam carriage** built by Nicholas Cugnot, a French officer. It was not very successful or efficient. The carriage had to be stopped every 15 minutes to build up the steam again (see page 2).

1803 The British engineer, Richard Trevithick, built the **first steam passenger vehicle** carrying eight passengers.

1859 An **internal combustion engine** that used gas as fuel was built by a Belgian, Jean-Joseph Lenoir.

1864 Siegfried Markus invented a motorized cart that used **petrol vapour**. His electrical system solved the problem of **ignition** (see page 26).

1876 German engineer, Nikolaus Otto, designed an engine which used the **four stroke engine cycle** (see page 10).

1885 The first car to run successfully with a **petrol engine** was built by German engineer, Karl Benz. This incorporated the internal combustion engine (see page 10).

1888 John Dunlop, from Ireland, made the **first *pneumatic* * tyre** (a hollow tyre filled with air). This was the main British contribution to the development of the motor car, but these tyres were first used in France.

1895 The **first motor car race** took place in France (see page 18). The course ran from Paris to Bordeaux and back.

1896 The **first four cylinder car engine** was built by René Panhard and Emile Levassor in France.

1898 The **first fully enclosed car**, a Renault, was produced in France.

1902 The French government tried to make motorists use **alchohol fuel** manufactured from potatoes.

1903 Vehicle registration plates were introduced in Britain.

1904 The land speed record of **100mph (161km/h)** was officially reached for the first time.

1905 Car bumpers were patented.

1906 The **first car racing track** was built at Brooklands in England.

1912 A car with an **electric self-starter** was introduced by Cadillac.

In **1913** Ford built the **first motor car assembly line**. The first car to be made was the **Ford Model T**. Until 1908, all cars had been hand made and were very expensive. But mass production greatly reduced the cost and time involved in making cars, enabling many more people to buy a car for the first time.

1916 Mechanical windscreen wipers were introduced.

1919 Traffic lights first appeared in the USA. Britain introduced them in **1928**.

1922 Ford became the **first firm to build over a million cars in one year**. This was matched by Volkswagen in **1962** and by British Leyland in **1968**.

1927 The Bugatti Royale, the **largest production car of all time**, was built in Italy. It was over 6.7m (22ft) long, with a 2m (7ft) bonnet, and a 4.3m (14ft 2 in) wheelbase. The 13-litre engine was so powerful that gears were not really needed.

1934 In France, Citroen introduced **front-wheel drive and unit construction** (with body and chassis in one unit). The **first drive-in cinema** also opened in New Jersey, USA.

1935 The **first person to exceed 300mph (483km/h)** on land was Malcom Campbell in the *Bluebird*.

1936 The **first diesel-engined private car**, a Mercedes-Benz, was built.

1959 The **BMC Mini**, designed by Sir Alec Issigonis, first sold for about £496 ($744). A total of 5.26 million were produced, giving Britain the lead in the world's small car market.

1971 An electric moon buggy was the **first car to be driven in Space**. Astronauts drove it on the Moon and left it there when they returned.

1983 The **world speed record** was broken by Richard Noble in *Thrust 2*. He sped across the Nevada Desert at **1019km/h (633 mph)**.

1992 Martin Brundle achieved the **highest speed ever reached by a standard production car** at the Nardo test track in Italy. He drove at a speed of 349.21km/h (217.1mph) in a Jaguar XJ220.

AMAZING CARS

*An **orange bubble car** was built in the 1970s by a fruit company to use as a publicity stunt to advertise their oranges.

***Two cars in one**. One inventor joined two "half cars" together, to provide an engine and controls at both ends. If two drivers worked very carefully together they could drive the car sideways.

*A **swan car** was built in 1912 by the Brooke Car Company. The car was in the shape of a swan at the front and was made purely for fun.

*The **longest car** in the world is over 30m (100ft) long. It was designed by Jay Ohrberg from the USA. This limousine has 26 wheels, a swimming pool and a water bed.

***Waterproof cars** are called Amphicars. They were built in the 1960s to travel on land and water.

*Several inventors have experimented with **flying cars**. One flying car was made especially for the James Bond film, *The Man With the Golden Gun*.

*The **most expensive British standard car** is the XJ220 Jaguar quoted at £40,2418 ($6, 018, 627).

FASCINATING FIGURES

Top 5 car producers

Japan	9,753,000 cars per year
USA	5,440,000
Germany	4,700,000
France	3,188,000
Spain	1,750,000

Did You Know?

*In six years, an average car travels nearly 97,000km (60, 276 miles). Each wheel will have turned 54 million times.
*The number of vehicles built in the world in 1994 was 49, 967 594, of which 36, 070 238 were cars.

ANSWERS TO QUESTIONS

Page 24 Dr Ferdinand Porsche designed the famous Volkswagen Beetle before he designed the Porsche 911.

Page 27 The sports car's front lights pop up so that they are high enough off the ground to give a good beam and be clearly seen by other cars. When the lights are not being used, they pop down into the body making the car more *streamlined**.

CUTAWAY PLANES

Clive Gifford

Designed by Steve Page

Illustrated by: Mark Franklin, Sean Wilkinson, Ian Cleaver, Robert Walster, and artists from the School of Illustration, Bournemouth and Poole College of Art and Design.

Usborne Publishing wish to thank the following for their help with this part of the book:

Airbus Industrie · Aircraft Research Association · Air International · Trevor Alner · British Aerospace Flying College Ltd. · Les Coombs · CSE Aviation · David Ditcher · Evans and Sutherland Computer Corporation · Flight Refuelling Ltd. · Helmet Integrated Systems · Hughes Corporation · Irvin GB Ltd. · Stephen Lake · Anthony Lawrence · Lord Corporation · Magellan Systems · Martin Baker Aircraft Company Ltd. · Martin Cross · Lockheed Martin Corporation · Michael Leek · Michelin · National Aerospace Laboratory (NLR) · Quadrant Picture Library · Stuart Priest · Richard Goode Aerobatic Displays · RFD Ltd. · Rolls-Royce Commercial Aero Engines Limited · Chris Sargent · SAS Flight Academy · Sextant Avionique · SFIM Industries · Clive Thomas · Thomson Training & Simulation · Steve Upson · Westland Aerospace

Contents

Introduction 34

Controlling a plane 36

Cockpits 38

Airliners 40

Engines 42

Aircraft construction 44

Aerodynamics and fuel 46

Learning to fly 48

Take-off and landing 50

VSTOL planes 52

Safety 54

Modern military planes 56

Navigation 58

Stealth planes 60

The future 62

Glossary 63

Planes: Facts 64

Words in *italic* type
Words in this section which appear in *italic* type and are followed by a small star (for example, *friction**) can be found in the glossary on page 63.

Introduction

The first successful flights people ever made were with the help of balloons filled with gases to make them lighter than air. It was not until 1903 that two Americans, the Wright Brothers, found a way to fly a craft that was heavier than air. In the time since the Wright Brothers' first flight, many advances and inventions have made modern planes look very different from early ones.

BAe Hawk 200

This is a single-seat jet fighter designed by computer with many of the latest advances in aircraft design. Its top speed is almost 1,045km/h (650mph) and when it is empty it weighs only a little more than four family cars.

The body of a plane is called the fuselage.

These parts of the tail help to keep the plane stable.

The Hawk's single jet engine is tucked into the back of the fuselage.

This is an external fuel tank. It holds more than eight times the fuel of a family car.

The Hawk 200's wingtips each hold a weapon called a Sidewinder missile.

You can see how the wing is made up of lots of strips all joined together. This makes the wing strong but much lighter than if it were made out of solid metal.

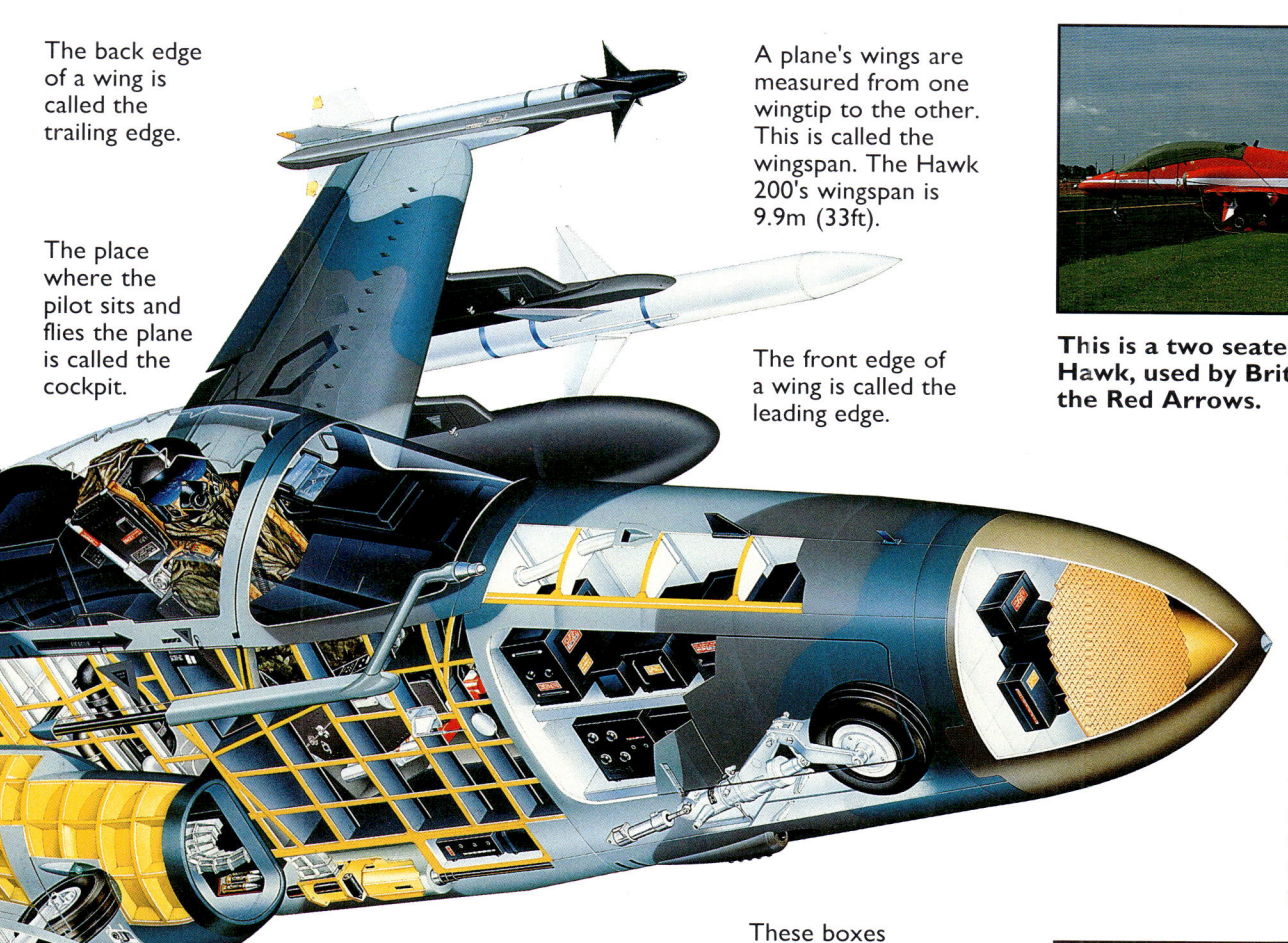

The back edge of a wing is called the trailing edge.

The place where the pilot sits and flies the plane is called the cockpit.

A plane's wings are measured from one wingtip to the other. This is called the wingspan. The Hawk 200's wingspan is 9.9m (33ft).

The front edge of a wing is called the leading edge.

This is a two seater version of the Hawk, used by British display team, the Red Arrows.

The nose of the Hawk contains advanced radar equipment.

This nosewheel and the wheels at the back are together called the landing gear.

This is an air intake. It channels air into the plane's jet engine. You can learn how a jet engine works on pages 42-43.

These boxes contain complicated electronics which help fly the aircraft. They are called avionics.

Wings

Wings are vital to planes. They provide the lift which takes the aircraft off the ground and into the air. They do this because of their specially curved shape.

As the engines pull the plane through the air, the leading edge of the wing divides the air, forcing some under it and some over. Both the top and bottom surfaces of the wing are curved but the top curves more steeply. The air moving over the wing, therefore has farther to go to catch up with the air moving underneath.

Slower-moving air presses more on the bottom of the wing than the faster moving air passing over the top of it. This stronger *air pressure** under the wing lifts it up. You can learn about car wings on page 8 and boat wings on page 85.

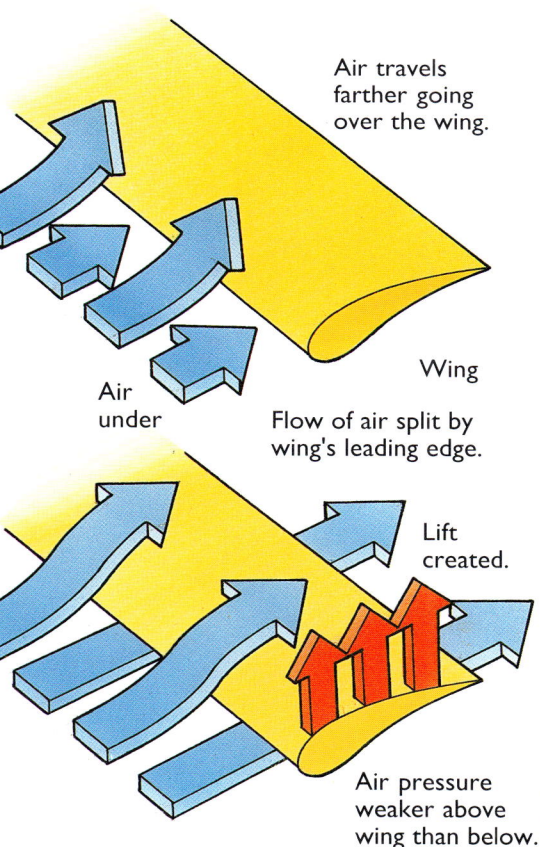

Air travels farther going over the wing.

Air under

Wing

Flow of air split by wing's leading edge.

Lift created.

Air pressure weaker above wing than below.

Wing types

There are many different wing shapes and sizes. Each type is best for a particular kind of plane. Here are three common wing types.

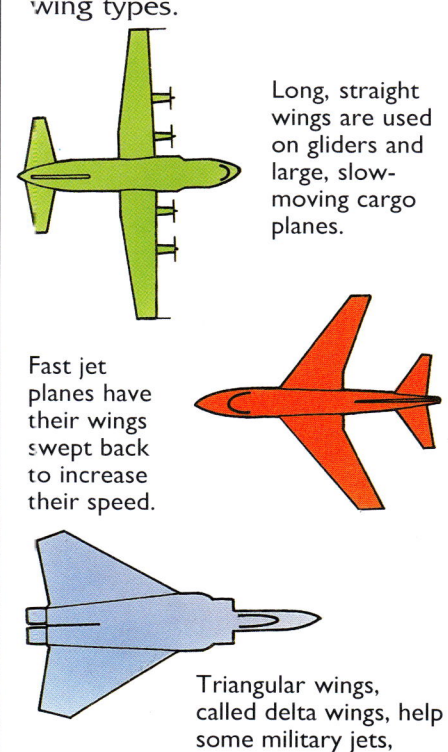

Long, straight wings are used on gliders and large, slow-moving cargo planes.

Fast jet planes have their wings swept back to increase their speed.

Triangular wings, called delta wings, help some military jets, travel even faster.

35

Controlling a plane

On the previous pages you saw how wings create lift. Lift is one of the *four forces of flight**. The others are thrust, drag and gravity. You can think of these four forces in pairs. For a plane to fly, lift must be greater than gravity, and thrust must be greater than drag.

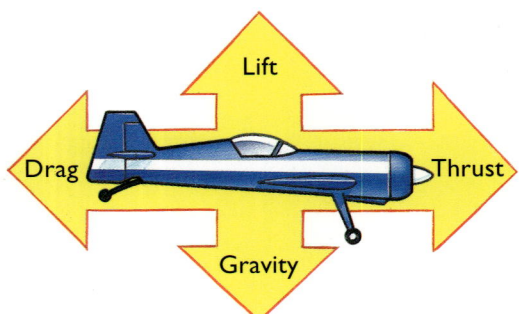

An aircraft uses its engines to create thrust and its wings to create lift. Modern planes are carefully designed to have less drag (see pages 46-47).

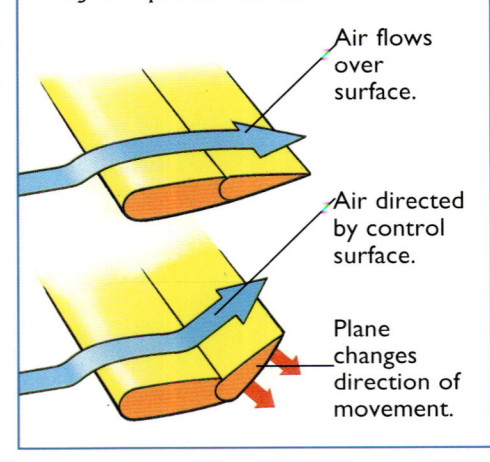

This cutaway picture of a Slingsby Firefly shows the control surfaces and how they are linked to the controls in the cockpit.

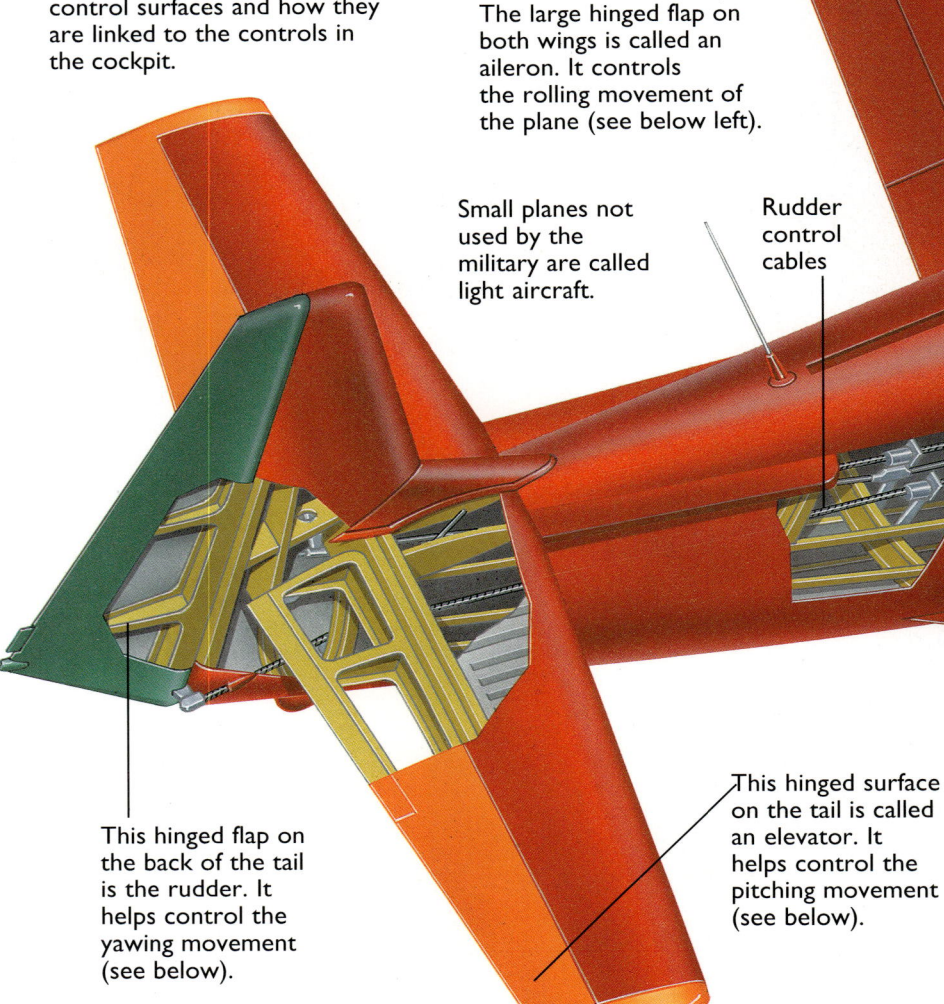

The large hinged flap on both wings is called an aileron. It controls the rolling movement of the plane (see below left).

Small planes not used by the military are called light aircraft.

Rudder control cables

This hinged flap on the back of the tail is the rudder. It helps control the yawing movement (see below).

This hinged surface on the tail is called an elevator. It helps control the pitching movement (see below).

Control surfaces

Once in the air, a plane must be controlled otherwise it will crash. The plane must be kept balanced but also be able to change direction.

Changing direction is done by moving hinged parts of the wing, called ailerons, and hinged parts of the tail called the elevators and rudder. Together, these are known as control surfaces. Control surfaces change the direction of some of the air flowing around the plane. The change in the airflow changes the way the plane moves.

- Air flows over surface.
- Air directed by control surface.
- Plane changes direction of movement.

Roll

Roll is when you move the wings up and down using the ailerons. For example, turning the left aileron down will push the left wing up, pushing the right wing down.

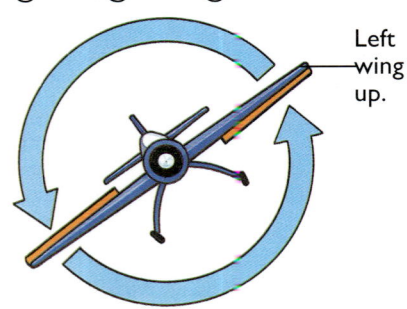

Left wing up.

Pitch

Pitch is when you move the fuselage up and down using the elevators. For example, turning the elevators down, lifts the tail up, causing the plane to dive down.

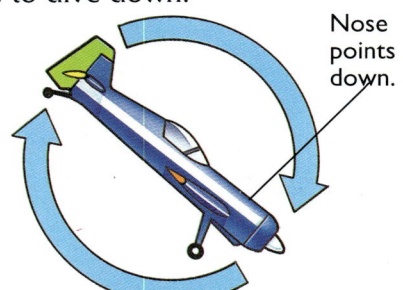

Nose points down.

Yaw

Yaw is when the plane moves from side to side. It is controlled by using the rudder. For example, turning the rudder left, will turn the plane's nose to the left.

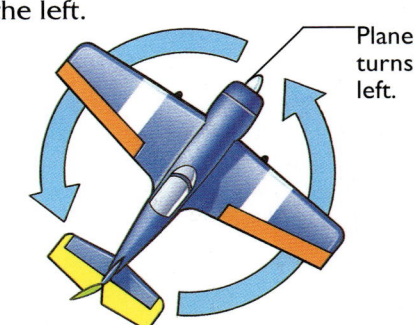

Plane turns left.

The Slingsby Firefly is made of modern materials. The rudder, for example, is made of plastic reinforced, or strengthened, with fine strands of glass.

This control column works two sets of control surfaces. Pushing it to the right and left moves the ailerons. Pushing it back and forth moves the elevators.

These foot plates push the rudder left or right. They are called rudder pedals.

This pulley allows the control wires to turn corners so they can be linked back to the controls in the cockpit.

This large hinged surface on the back of the wing is called a flap. It is usually used for take-off and landing.

Flaps extend out, to make the wing bigger. This creates more lift but also more drag which slows the plane down.

Banked turns

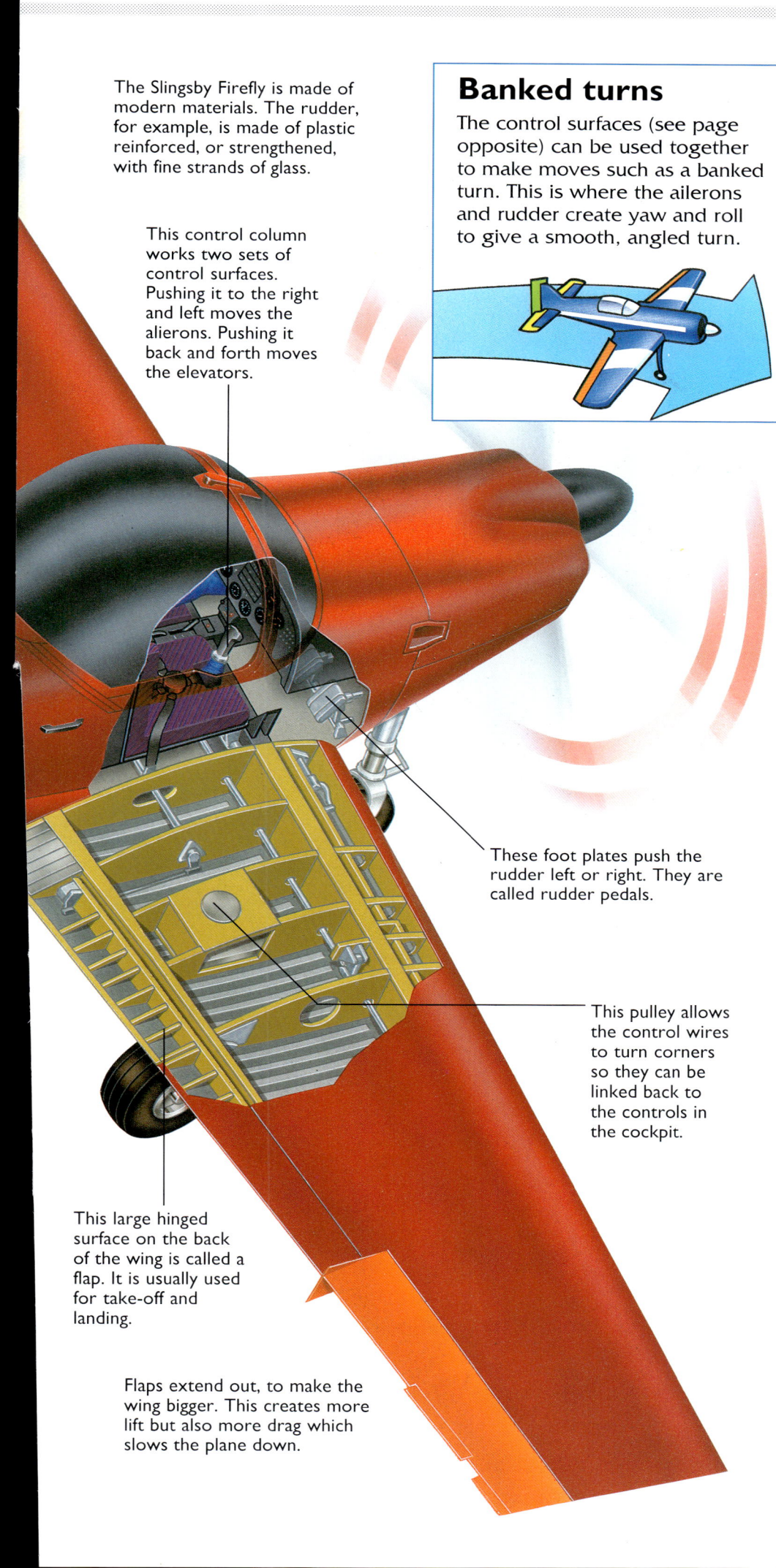

The control surfaces (see page opposite) can be used together to make moves such as a banked turn. This is where the ailerons and rudder create yaw and roll to give a smooth, angled turn.

Stalling

When a plane climbs, it is angled up and the air flows less smoothly over the wings. If an aircraft tries to climb too steeply without enough power, there is not enough speed for the air to flow over the wings and create lift. The flow of air breaks up and the aircraft begins to fall out of the sky. This is called *stalling*.

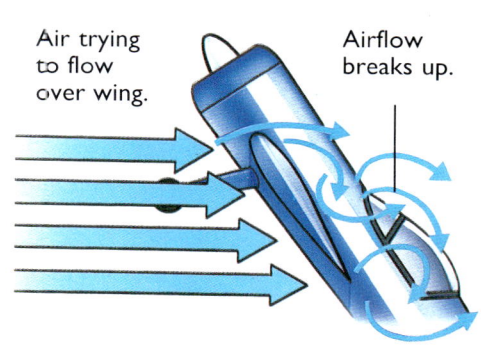

Air trying to flow over wing.

Airflow breaks up.

Pilots are taught how not to stall and how to recover if they do, but it can still be dangerous if the plane is near the ground. Modern planes have a complex set of sensors to help prevent stalling which is called a stall warning system.

Other surfaces

In addition to elevators, ailerons and a rudder, most modern planes have other control surfaces such as spoilers and slats. Slats do a similar job to flaps (see labels just to the left). They extend forward from the front, or leading edge, of the wing. Spoilers are large panels built into the wing which can lift up and 'spoil' the flow of air over the wing. They create less lift and more drag. They are used either to increase the effect of the ailerons or as air brakes (see page 51) to slow the aircraft down.

This airliner wing has its control surfaces arranged for just before landing.

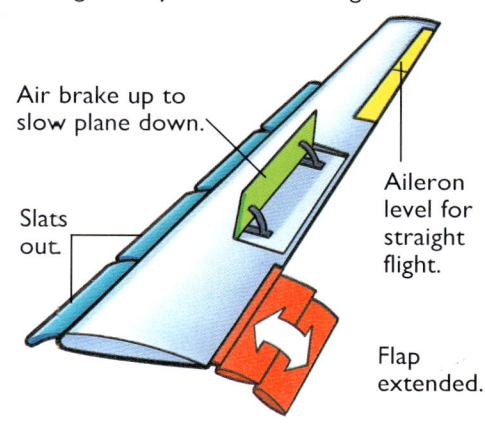

Air brake up to slow plane down.

Slats out.

Aileron level for straight flight.

Flap extended.

Cockpits

The cockpit is where the pilot sits and controls the aircraft. Instruments show how various parts of the plane are performing while navigation systems such as radar (see page 58) and the artificial horizon help keep it on the correct route.

The Optica's cockpit gives a clear view all the way around. It is ideal for aerial observation and photography.

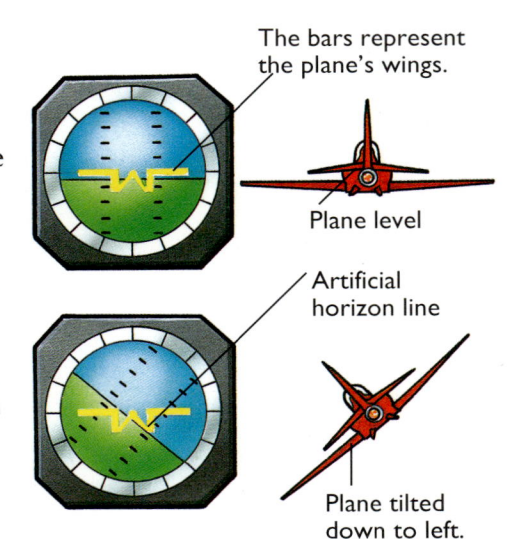

Early cockpits

Early cockpits were often very uncomfortable. Many were open-air and left the pilot and any passengers unprotected from the weather.

A modern aircraft cockpit is full of controls and instruments, but the cockpits of the first planes were empty in comparison. You can see how simple an early aircraft's cockpit is by looking at the one in a 1933 De Havilland Leopard Moth (see right).

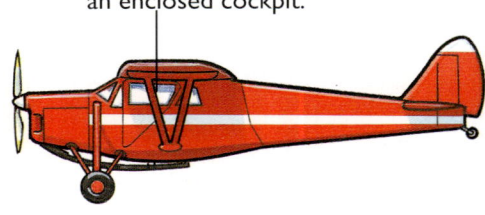

The Leopard Moth did have an enclosed cockpit.

There were 133 De Havilland Leopard Moths built in the 1930s.

From cockpit to wing

In gliders and light aircraft, the cockpit controls are linked directly to the *control surfaces** by a system of cables and pulleys.

In bigger aircraft, they have *hydraulic** systems instead of mechanical cables. Hydraulic power is created by putting a liquid under great pressure, as the diagram on the right shows. A liquid which doesn't freeze, even in very cold weather, is used to fill the cylinders and pipes of a hydraulic system.

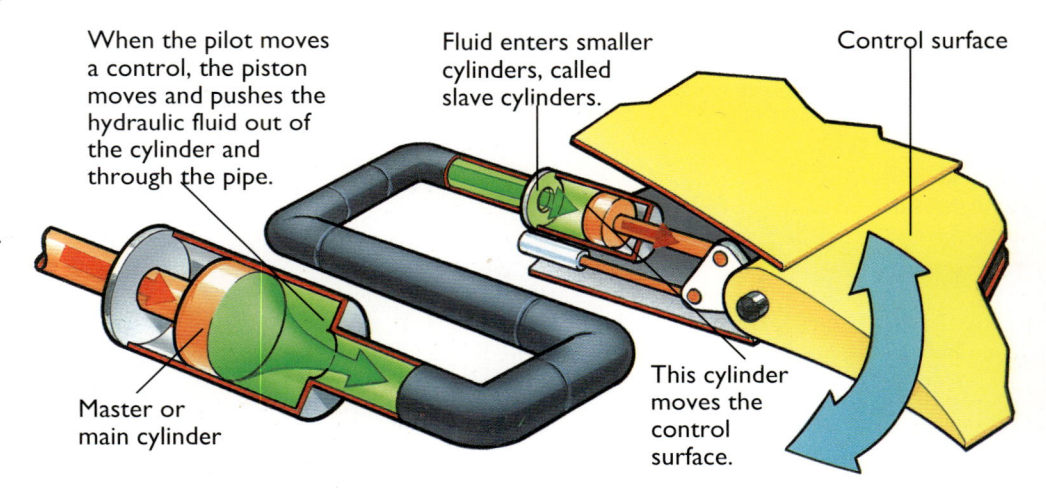

Artificial Horizon

The artificial horizon tells the pilot if the plane is flying level. A gyroscope (see page 58) keeps a line on the dial exactly parallel with the horizon of the Earth. This line is the artificial horizon. When the aircraft tips one way or the other, the markings that show the plane's wings dip to either side of the horizon line.

For example, when the aircraft does a banked turn (see page 37) the horizon line appears to tilt. In fact, it remains parallel with the horizon of the Earth and the rest of the plane (including the wings on the dial) tilt around it.

The bars represent the plane's wings.
Plane level
Artificial horizon line
Plane tilted down to left.

Thermometer
Altimeter (see right).
This is the plane's airspeed indicator (see right).
Control column

When the pilot moves a control, the piston moves and pushes the hydraulic fluid out of the cylinder and through the pipe.

Fluid enters smaller cylinders, called slave cylinders.

Control surface

Master or main cylinder

This cylinder moves the control surface.

38

Pitot tube and airspeed indicator

The pitot tube is a small tube mounted on the wing or body of the aircraft. It measures two types of *air pressure**: the pressure of the air around it, called static air pressure, and the pressure of the air rushing at it as the plane flies, called ram pressure. This information is fed into the aircraft's instruments.

A plane's speed through the air is shown by the airspeed indicator, or ASI. The relationship between the two air pressure readings from the pitot tube is changed by the ASI into the airspeed of the aircraft.

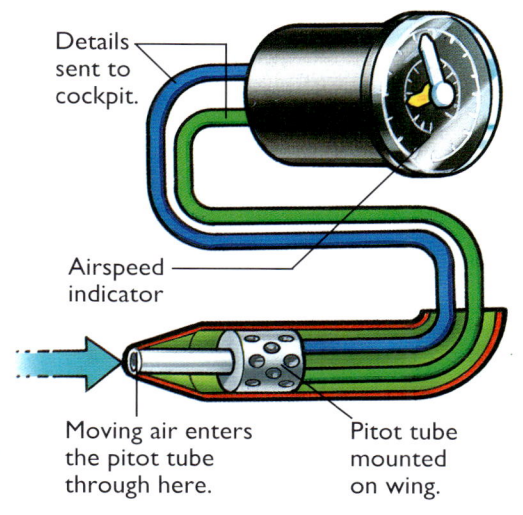

Details sent to cockpit.
Airspeed indicator
Moving air enters the pitot tube through here.
Pitot tube mounted on wing.

Altimeter

The altimeter tells the pilot the altitude, or height above ground, of the plane. Some altimeters use radar (see page 58) but others use the static pressure reading from the pitot tube. *Air pressure** gets less the higher you fly, so the altimeter can calculate the plane's height from the surrounding air pressure.

This sealed capsule of air helps the altimeter make its reading.

Modern cockpits

As aircraft became more advanced, the numbers of controls and instruments increased and cockpits became much more complicated.

Advanced airliners such as the Boeing 777 and Airbus A340 have their cockpits, called flight decks, specially designed to be simpler and better laid out than before. The cockpit of the A340 (shown on the left) has computer screens which display different types of information, instead of rows of individual instruments.

Pilot's seat
The throttles control the speed of the engines.
Fly by wire joystick (see below).
Co-pilot's seat

Here is a test model of the A340 airliner on the ground at the Farnborough Airshow in England.

Fly-by-wire

In the very latest aircraft, some hydraulic systems are replaced with an advanced system called fly-by-wire. With fly-by-wire, electric wires run from a computer to a mixture of electric motors and small hydraulic systems called *actuators** in the wings and tail. When the pilot moves the controls, signals from the computer instruct the motors and hydraulics to move the control surfaces. Fly-by-wire is very precise, much lighter and easier to repair than hydraulics and works well with the plane's flight computers.

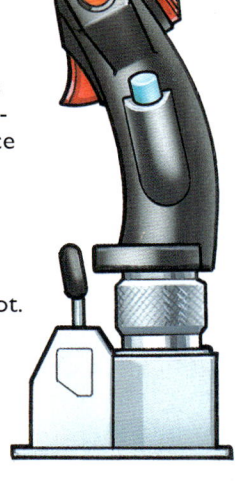

Many cockpits which have fly-by-wire replace the control column with this type of small joystick placed to one side of the pilot.

The General Dynamics F16 Fighting Falcon was one of the first military jets to feature fly-by-wire controls.

Airliners

Airliners are planes that carry large numbers of passengers. The earliest airliners carried just a few people and were cold and noisy. These planes could only travel long distances by making small hops from place to place.

This Ford Tri-motor had a maximum range of just over 800km (500 miles). Some of today's airliners can fly 20 times as far without having to refuel.

Airbus A340-300 and -200

The A340 is the latest in a line of airliners built by several European companies working together under the name Airbus Industrie. There are two versions of the A340. The -300 (shown below) has a long *range* * but the -200 can travel even farther. The -200's fuselage is 4m (13ft) shorter and the plane has fewer seats but more fuel. In 1993, an Airbus A340-200 flew around the world stopping only once, in New Zealand, to refuel. The whole journey took just over 48 hours.

Here's an Airbus A340-300 test aircraft in flight.

Airliner fuselages must be incredibly strong to withstand the pressure of high altitude flight and protect the passengers and crew inside.

This is one of the A340's eight passenger exits.

Here is the inner shell of the airliner's fuselage.

Pressurization

As a plane flies higher and higher, the air around it gets thinner. At the altitudes a modern jet airliner flies, there is not enough air for the passengers to breathe. What they do is to keep the airliner's airtight fuselage filled with *pressurized* * air that the passengers and crew can breathe normally.

On the ground the pressure of the air inside the plane is the same as the air outside it.

The top floor or deck holds the passengers. It is called the cabin.

Here you can see the complicated structure of the wing.

These air brakes are positioned up to slow the airliner down. You can learn more about air brakes on page 51.

At great heights, the air has less pressure and is much thinner.

In flight, the air inside the plane is kept to a similar pressure to the air on the ground.

Wing rib

Because of concerns about the environment, new airliners' engines have to be cleaner and quieter than before.

Fan blades

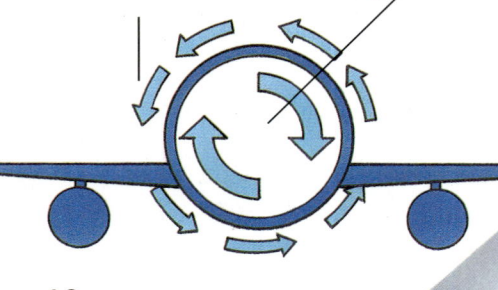

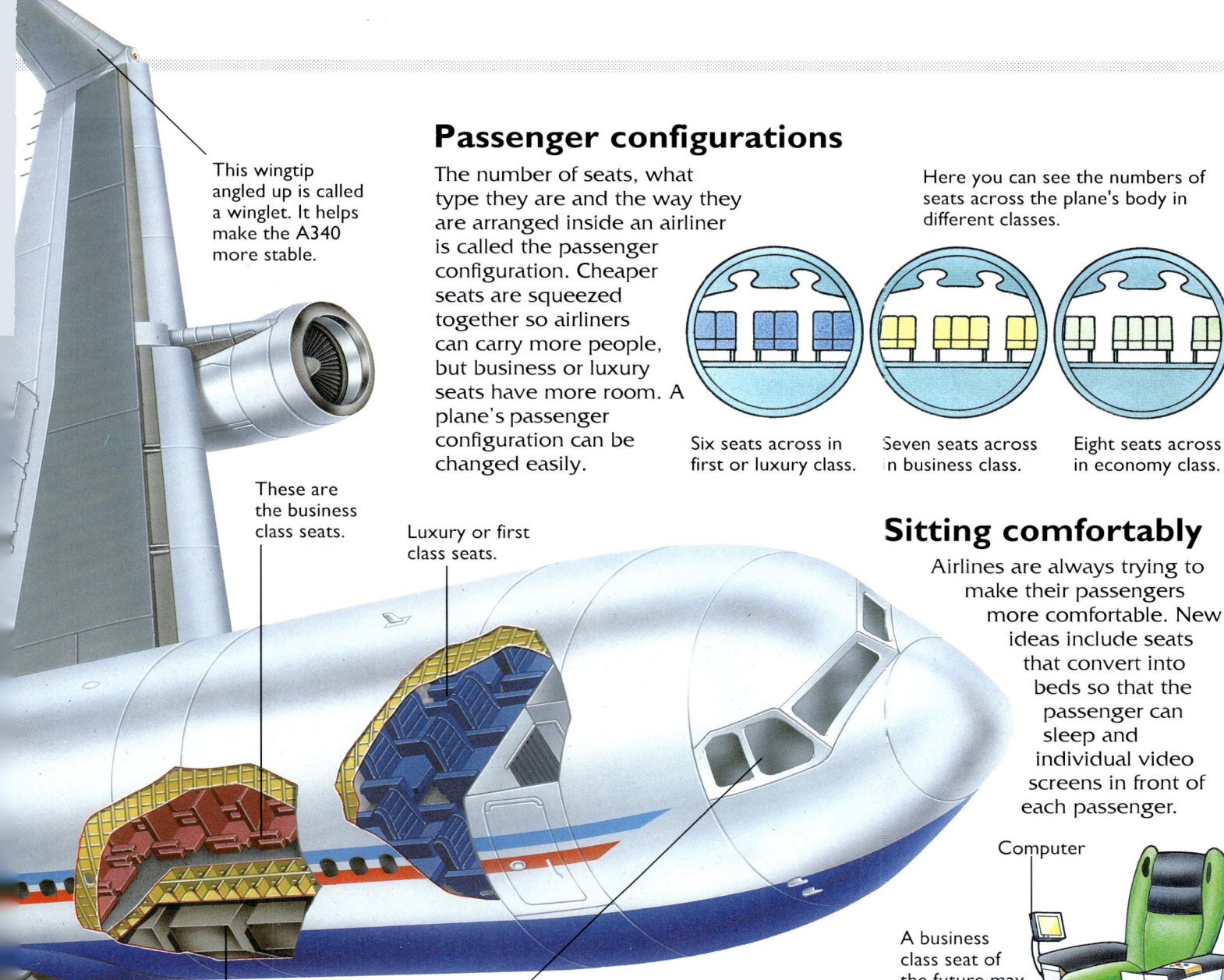

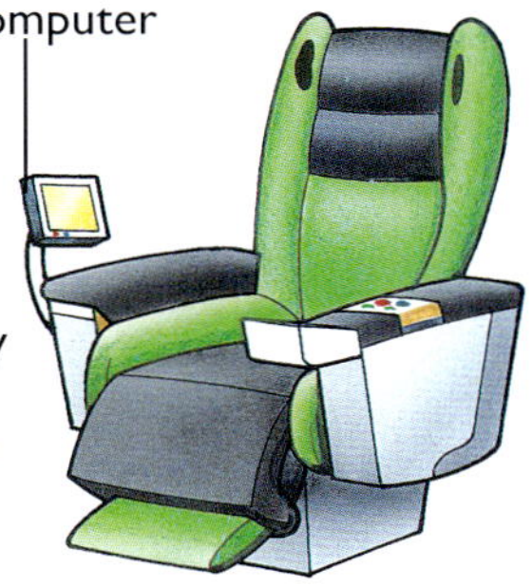

This wingtip angled up is called a winglet. It helps make the A340 more stable.

Passenger configurations

The number of seats, what type they are and the way they are arranged inside an airliner is called the passenger configuration. Cheaper seats are squeezed together so airliners can carry more people, but business or luxury seats have more room. A plane's passenger configuration can be changed easily.

Here you can see the numbers of seats across the plane's body in different classes.

Six seats across in first or luxury class.

Seven seats across in business class.

Eight seats across in economy class.

These are the business class seats.

Luxury or first class seats.

Sitting comfortably

Airlines are always trying to make their passengers more comfortable. New ideas include seats that convert into beds so that the passenger can sleep and individual video screens in front of each passenger.

Computer

A business class seat of the future may have its own computer and fax machine.

The lower floor is the luggage or cargo hold.

You can see the A340's flight deck on page 39.

All the other staff in the plane, apart from the pilot, are called aircrew.

Loading and unloading

All airliners carry some cargo, from the passenger's suitcases to mail and packages. When they land, many airliners have to unload the passengers, crew and cargo and also be checked, filled with fuel and then loaded again for another journey. This is called turning around an aircraft and it involves many people working together.

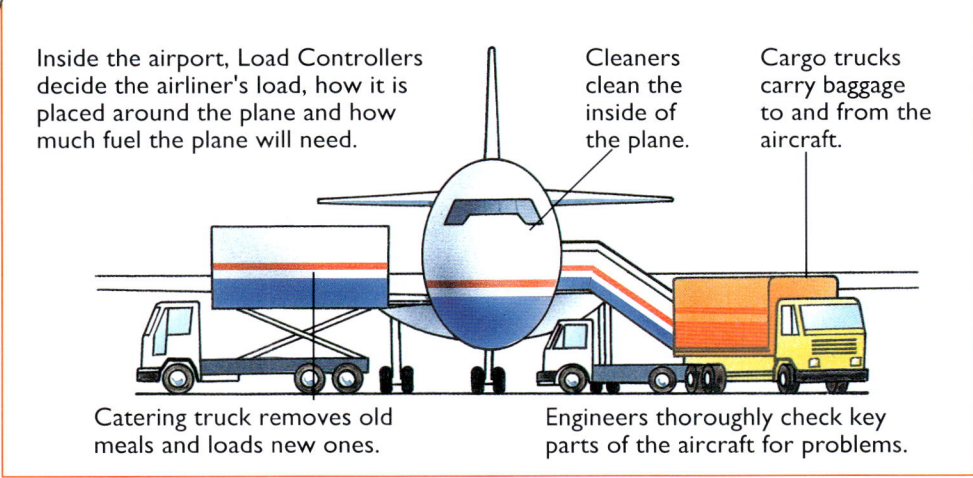

Inside the airport, Load Controllers decide the airliner's load, how it is placed around the plane and how much fuel the plane will need.

Cleaners clean the inside of the plane.

Cargo trucks carry baggage to and from the aircraft.

Catering truck removes old meals and loads new ones.

Engineers thoroughly check key parts of the aircraft for problems.

Cargo planes

Some airliners have been rebuilt to carry just cargo. The latest large cargo carrier is a new version of the Airbus A300 airliner, called the A300-600ST. It can carry enormous pieces of machinery or rocket and plane parts up to a weight of 45 metric tonnes (50 tons).

The A300-600ST's fuselage is over 7.2m (24ft) across.

Engines

All aircraft need one or more engines to push them forward. Many use piston engines to drive propellers. Fast military planes and airliners use jet, or gas turbine engines.

Piston engines

The first piston engines were built for cars in the 1880s. Aircraft piston engines work in a similar way to those in cars (see pages 10-11). Instead of driving wheels around though, they spin a propeller at very high speeds. This propeller pulls the aircraft through the air. An aircraft engine is powered by a mixture of fuel and air injected into the engine's cylinders, where it is *ignited** and burned.

This diagram shows how the pistons move inside the cylinders of a piston engine.

1. Air and fuel are injected into the cylinder.

2. The piston moves up the cylinder, squeezing air and fuel together.

3. The mixture is burned to create gases which push the piston down the cylinder.

4. The piston moves up the cylinder and pushes out waste gases.

Crank converts up and down movement of piston into a turning movement.

Crank passes turning power via the gears to the propeller shaft.

Gas turbine engines

A gas turbine engine burns fuel and air, a little like a piston engine. Instead of moving a piston up and down, though, it drives a series of fan blades, called a turbine, around at high speeds. Gas turbine engines were first used to power aircraft in World War Two. There are many different types of gas turbine engine. The most common is the turbofan engine. These are found on modern jet airliners, military cargo transporters and many other types of jet aircraft.

Rolls Royce RB211-535

This is a powerful turbofan engine built by the British company, Rolls Royce. The power of an engine is measured in pounds or kN. The RB211-535 produces 192kN (43,100lb) of thrust, more power than is produced by 50 family cars. It is found on modern airliners such as the Russian Tupolev Tu204.

This Boeing 757 is powered by two RB211-535 engines.

This large intake fan is a very efficient propeller. It pushes huge amounts of air through the engine.

The fan is just over 2m (74in) across.

Each blade is made from a complex mixture of metals called an alloy.

Most of the air sucked in runs straight through the engine and out of the back to produce lots of thrust. Up to 75% of the engine's power is created in this way.

The rest of the air enters this compression chamber where it is squeezed together by small sets of fans called compressor blades.

The air is put under enormous pressure, over 30 times as much as normal.

This is the engine's bypass duct.

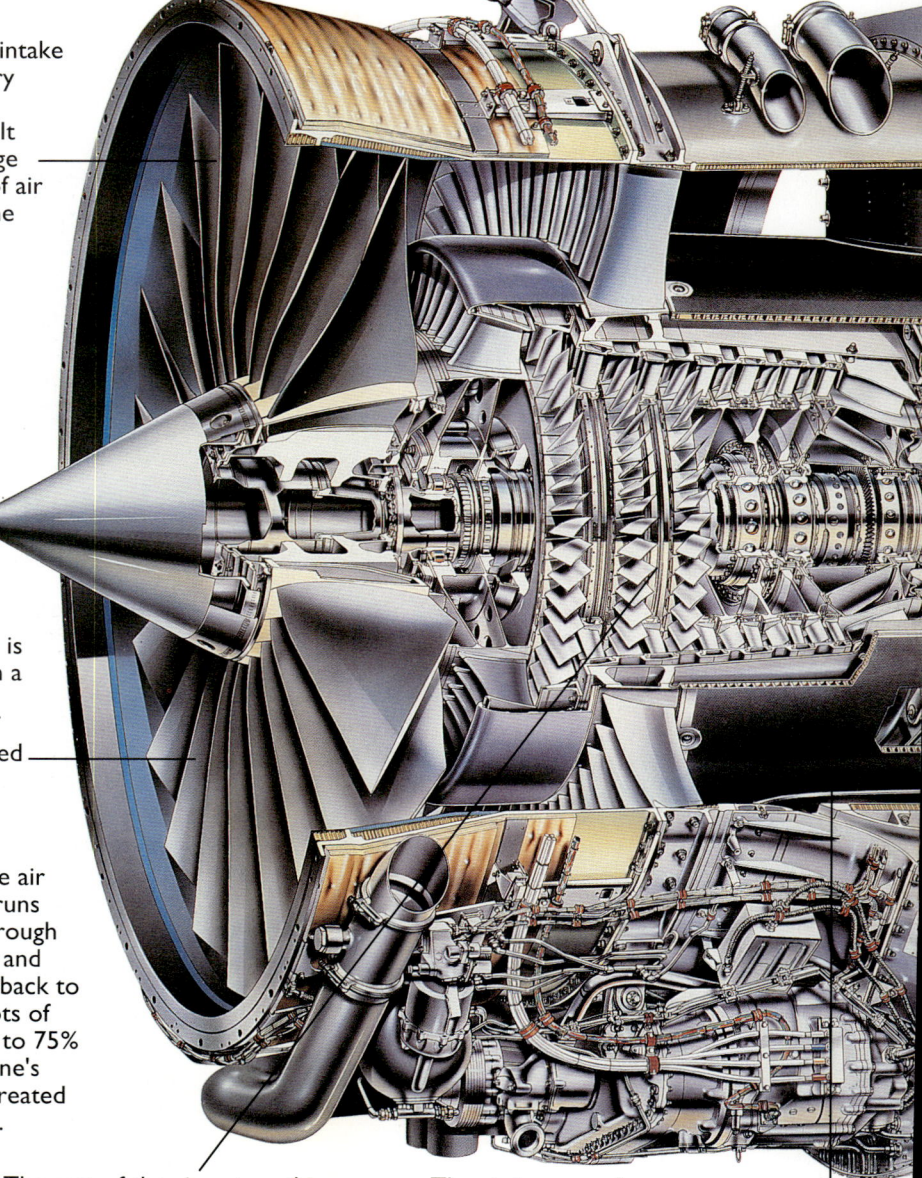

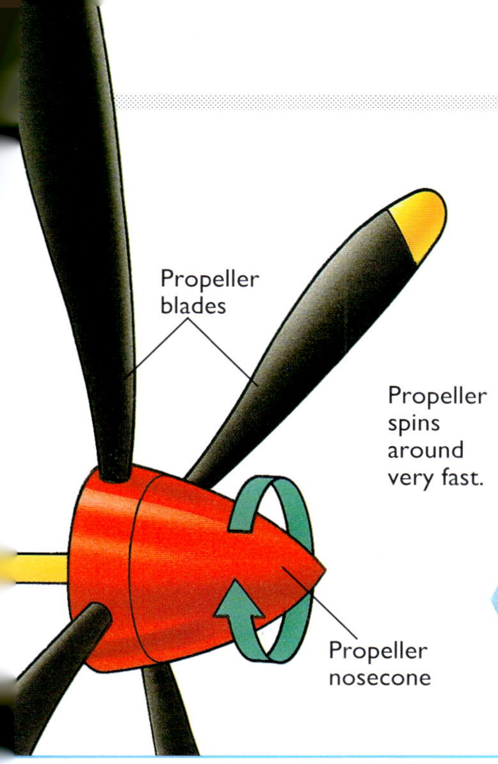

Propellers

Propellers are made up of between two and five blades. These are shaped to cut through the air and push it back. Pushing the air back, pulls the aircraft forward. The angle of each blade is called the pitch.

Propeller blades

Propeller spins around very fast.

Propeller nosecone

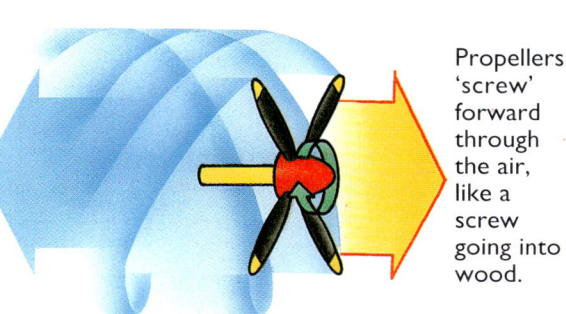

Propellers 'screw' forward through the air, like a screw going into wood.

Many propellers have variable pitch. This means that they can change the angle of their blades for different flying jobs. For example, a steep climb needs more pulling power while level cruising needs less.

Propellers can have different numbers of blades. This 12-bladed propeller is on a Russian Antonov An76 research aircraft.

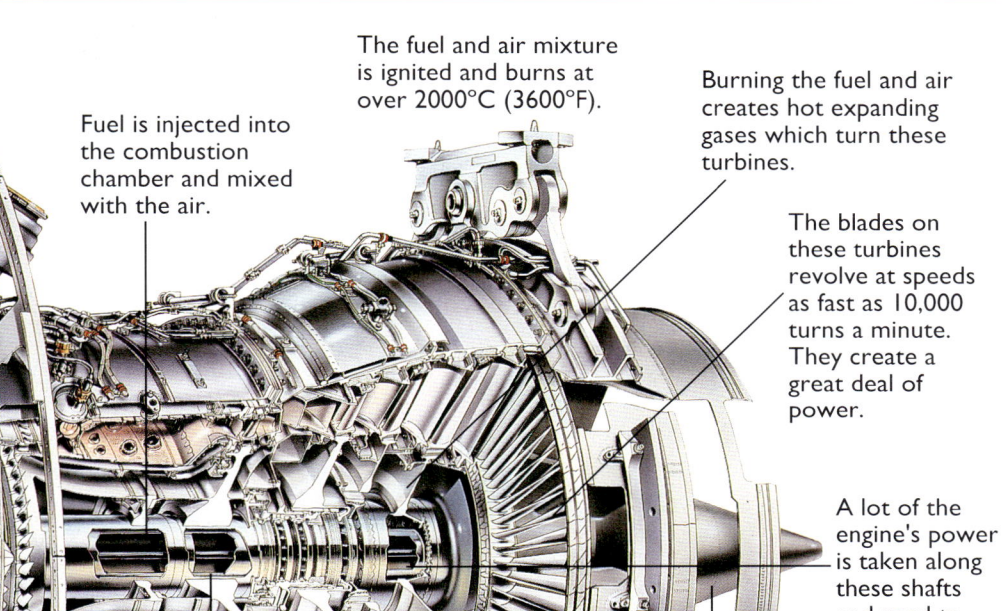

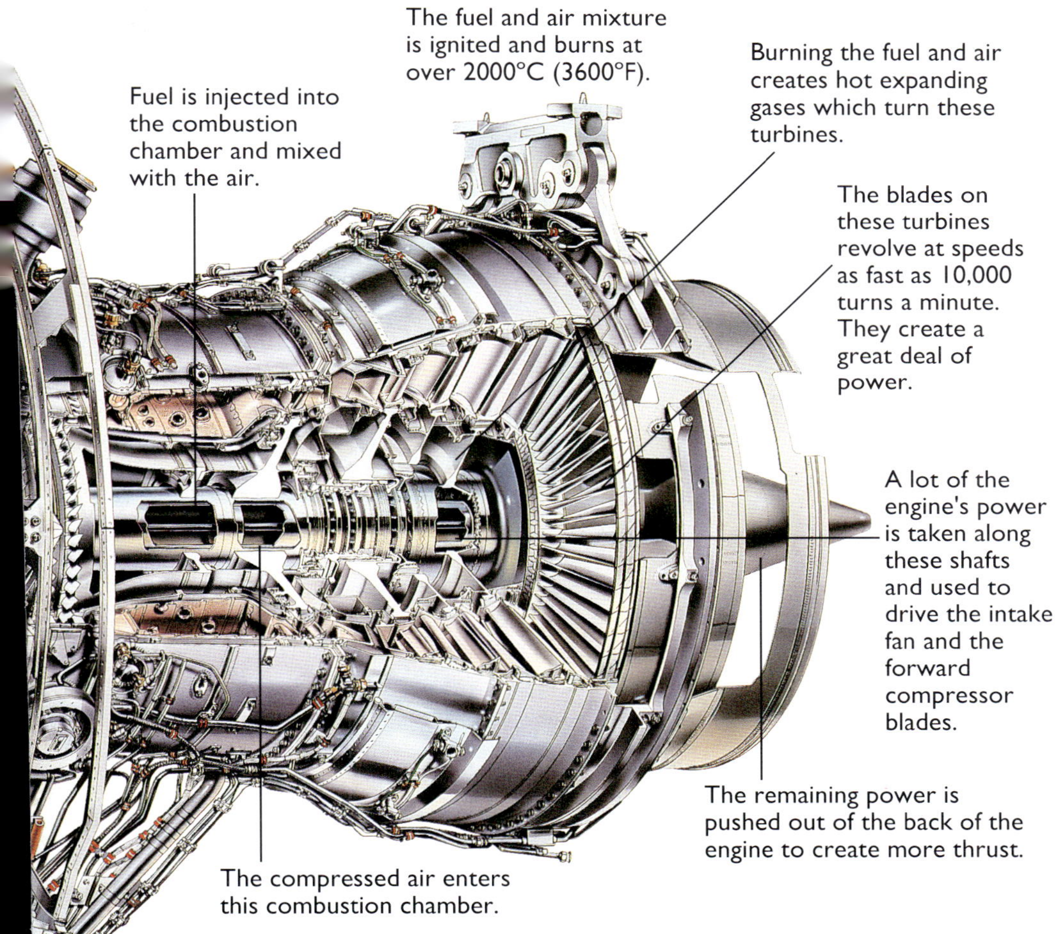

Fuel is injected into the combustion chamber and mixed with the air.

The fuel and air mixture is ignited and burns at over 2000°C (3600°F).

Burning the fuel and air creates hot expanding gases which turn these turbines.

The blades on these turbines revolve at speeds as fast as 10,000 turns a minute. They create a great deal of power.

A lot of the engine's power is taken along these shafts and used to drive the intake fan and the forward compressor blades.

The remaining power is pushed out of the back of the engine to create more thrust.

The compressed air enters this combustion chamber.

Splitting the air

Air sucked in at the front of a turbofan engine is split. Some enters the core of the engine where it helps burn fuel to produce thrust. However, as much as ten times that amount of air bypasses the middle and flows through the engine's bypass duct. This air creates much more thrust and also helps cool the engine core.

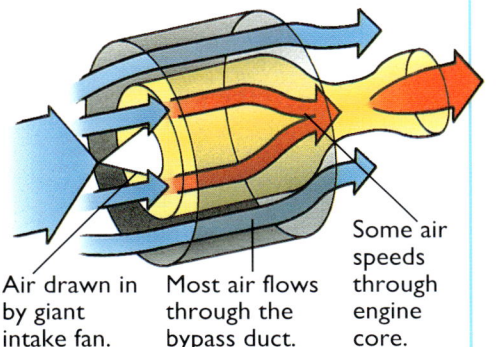

Air drawn in by giant intake fan.

Most air flows through the bypass duct.

Some air speeds through engine core.

Turbojets and props

Turbojets are the simplest type of gas turbine engine. Thrust only comes from the hot gases being pushed out of the back of the engine. Turbojets produce very high speeds but tend to be noisy and use a lot of fuel. They are found on some fast modern jet aircraft.

Concorde is powered by four Olympus 593 turbojet engines.

Turboprop engines have an extra turbine which uses much of the thrust created by the engine to drive a propeller. Turboprops burn less fuel and are much quieter than other turbine engines but cannot fly faster than about 800km/h (500mph).

This Fokker F27 has two turboprops.

Aircraft construction

The very first planes were built from materials that were easily available and were quite light. Certain types of light wood were used as a skeleton which was then covered with cloth stretched tight.

Although these planes were light enough to be lifted into the air by the weak engines they had at the time, they were not very strong.

LVG CVI

This German LVG CVI two-seat bomber was built near the end of World War One. By that time, advances in aircraft design had helped to make planes less flimsy. For its time, the LVG CVI was a strong plane, but heavy and quite slow.

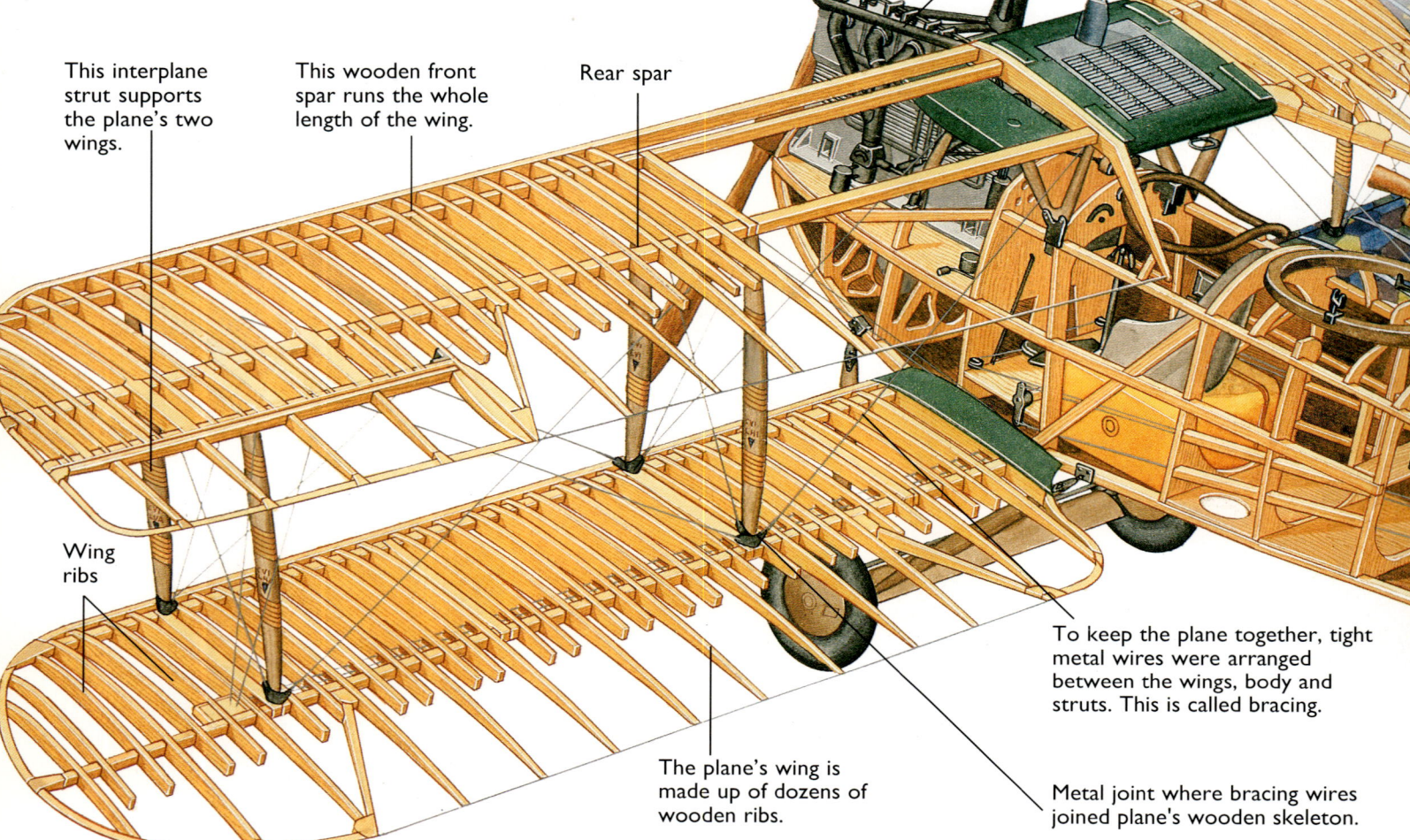

The engine produced almost as much power as three modern family car engines. Yet, because of its weight and shape, it's top speed was only 170km/h (105mph).

This interplane strut supports the plane's two wings.

This wooden front spar runs the whole length of the wing.

Rear spar

Wing ribs

The plane's wing is made up of dozens of wooden ribs.

To keep the plane together, tight metal wires were arranged between the wings, body and struts. This is called bracing.

Metal joint where bracing wires joined plane's wooden skeleton.

Advances in aircraft structure

After World War One, planes began to be built using more metal. Many plane makers used metal or thin sheets of wood instead of the outer covering of fabric. They also used tough metal steel tubing instead of wood for the inner skeleton.

The design of planes changed greatly as *duralumin**, a metal *alloy**, began to be made in sheets. Duralumin was used to make a stressed metal skin which was strong but lighter than large parts of the inner skeleton which it replaced.

Other metal alloys have since been invented which are stronger or lighter than duralumin, or are more heat resistant. Some of these contain a metal called titanium which is very strong and has an extremely high melting point.

Here you can see part of the steel tube framework of an American, Curtiss Hawk plane.

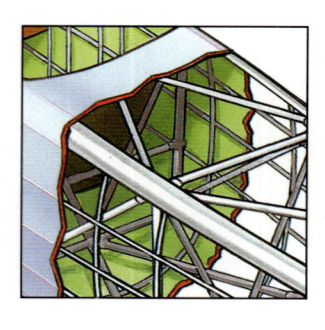

This German BF109G has a stressed skin shell called a monocoque.

Many parts of this Sukhoi Su26 are made from advanced metal alloys.

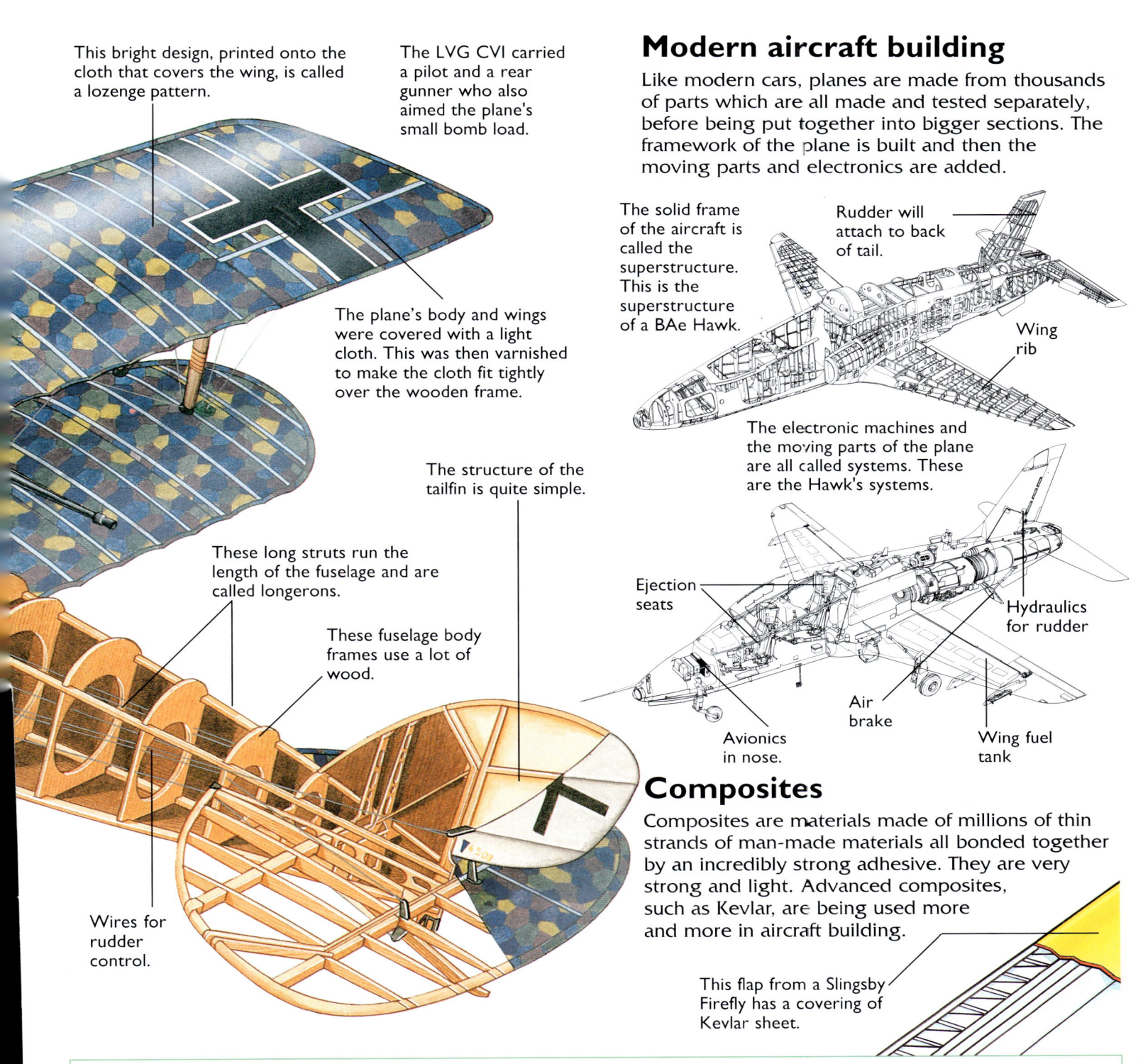

This bright design, printed onto the cloth that covers the wing, is called a lozenge pattern.

The LVG CVI carried a pilot and a rear gunner who also aimed the plane's small bomb load.

The plane's body and wings were covered with a light cloth. This was then varnished to make the cloth fit tightly over the wooden frame.

The structure of the tailfin is quite simple.

These long struts run the length of the fuselage and are called longerons.

These fuselage body frames use a lot of wood.

Wires for rudder control.

Modern aircraft building

Like modern cars, planes are made from thousands of parts which are all made and tested separately, before being put together into bigger sections. The framework of the plane is built and then the moving parts and electronics are added.

The solid frame of the aircraft is called the superstructure. This is the superstructure of a BAe Hawk.

Rudder will attach to back of tail.

Wing rib

The electronic machines and the moving parts of the plane are all called systems. These are the Hawk's systems.

Ejection seats

Hydraulics for rudder

Avionics in nose.

Air brake

Wing fuel tank

Composites

Composites are materials made of millions of thin strands of man-made materials all bonded together by an incredibly strong adhesive. They are very strong and light. Advanced composites, such as Kevlar, are being used more and more in aircraft building.

This flap from a Slingsby Firefly has a covering of Kevlar sheet.

Computer Aided Design

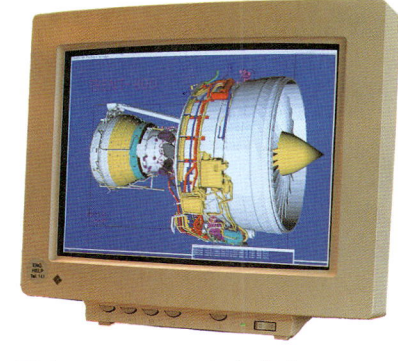

Computer Aided Design (C.A.D.) uses powerful computers that allow engineers to experiment with and accurately design aircraft parts on-screen. Once the design is decided upon, other computers simulate extreme conditions such as cold, heat, wear and stress, to test these parts thoroughly. C.A.D. is also used to design cars (see page 4).

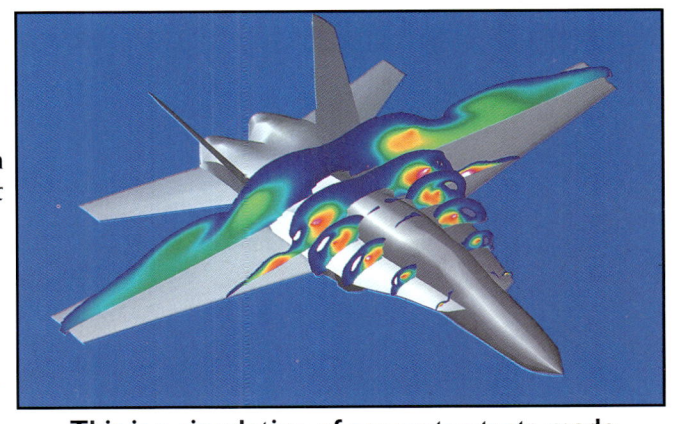

This computer is helping design a Rolls Royce engine.

This is a simulation of computer tests made on a military jet's structure.

Aerodynamics and fuel

Aerodynamics is the study of how a moving object travels through a liquid or a gas, such as air. It is vital to know how planes fly for lots of reasons. When a plane moves forward, the air flows over the wings which creates lift. When the air flows over the *control surfaces**, it allows the plane to change direction. When air flows over the whole plane, it creates drag. Drag slows a plane down and makes it use more fuel. Changing the shape and surface of a plane to reduce drag is called *streamlining**.

Box shape
Air flow

Air hits flat front of plane and cannot pass by easily. This creates lots of drag.

Air flows smoothly around a modern jet.

Concorde's delta wing shape was carefully designed to create the maximum amount of lift with as little drag as possible.

As many as 128 passenger seats can be squeezed close together.

Concorde

Concorde is the fastest airliner in the world with a top speed of 2300km/h (1430mph), more than twice the speed of other modern airliners. Concorde is so fast mainly because of its incredibly powerful engines and its advanced aerodynamic shape which is very streamlined.

This is the galley where food is prepared.

The pointed nose helps cut through the air.

The nose stays straight when flying but drops down to let the pilot see better for take-off and landing.

The flight deck seats three people.

Twin nosewheels

The fuselage is long and thin with a smooth, rounded shape. This makes it more streamlined.

Concorde has been carefully streamlined so that few parts stick out from its smooth shape.

Here you can see the complicated structure of the wing which is needed to make it strong while keeping it light.

Concorde's cruising speed means it flies 10km (6 miles) every 15 seconds.

Advances in streamlining

Early aircraft with lots of struts and wires made no real attempts at streamlining, but as planes got faster and exceeded 200km/h (125mph) streamlining became more important. Gradually, the number of struts and wires holding a plane together was reduced and the ones that were left were shaped to help the air flow more smoothly around them. Engines which had been left out in the open were later put inside smoothly-shaped engine covers which are called cowlings.

Engine cowling
Air flows smoothly past.

Monoplanes with one set of wings, which needed almost no struts or bracing, started to replace *biplanes** during the twenties and thirties. Fixed landing gears were replaced with retractable ones (see page 50). These could be lowered for take-off and landing and raised to create much less drag when flying.

The Supermarine Spitfire was among the first military planes to have a retractable landing gear.

With the arrival of jet engines, aircraft could travel much faster and this called for more changes in the shape of aircraft. Their surfaces were made as smooth as possible. The front of planes became sharper to cut through the air, and wings were swept back to make the aircraft even more streamlined.

The Hawker Hunter's swept back wings helped it reach a top speed of over 1100km/h (690mph).

46

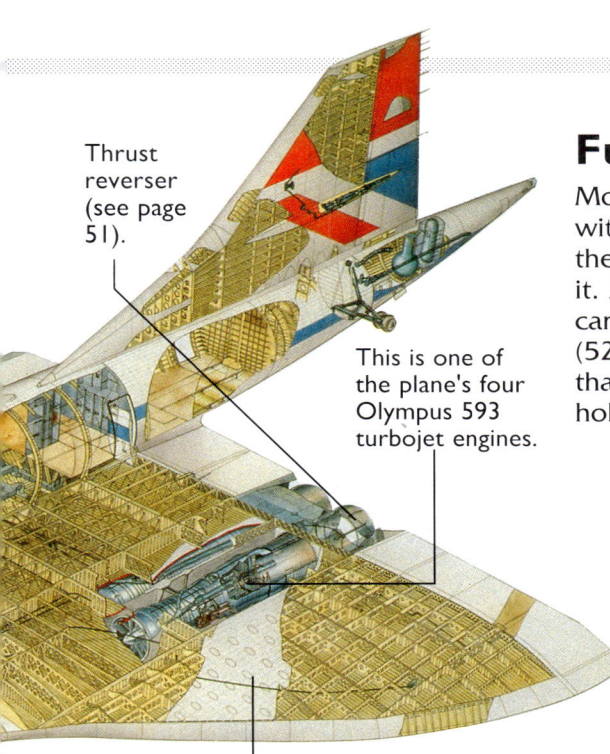

Thrust reverser (see page 51).

This is one of the plane's four Olympus 593 turbojet engines.

Concorde flies high where the air may be colder than -60°C (-76°F). These de-icing panels stop ice from forming on the wings which can reduce lift.

Fuel

Modern aircraft, especially those with jet engines, use fuel so quickly they have to carry huge amounts of it. A Boeing 747, for example, can carry as much as 197,000 litres (52,000 gallons) of fuel. Compare that to a typical family car which holds only 70 litres (15 gallons).

A plane must stay balanced as it flies. As fuel is used up, the balance of a plane changes. Many modern aircraft have *fuel management systems**. These measure how much fuel is in each tank and can switch which tank supplies fuel at any time, to even up their weight.

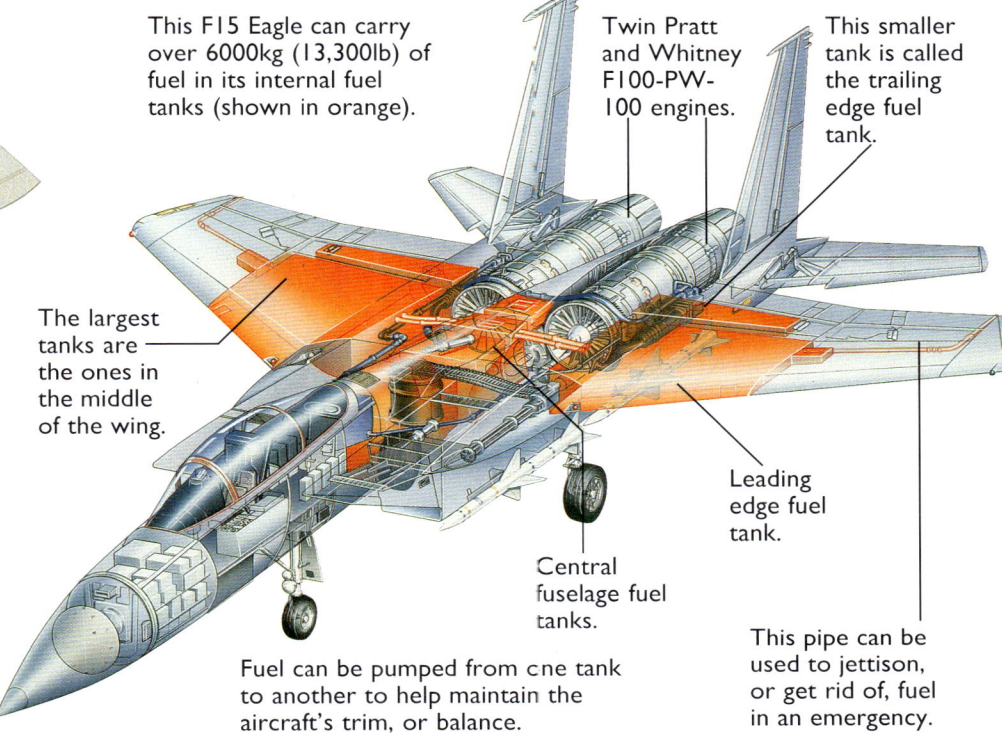

This F15 Eagle can carry over 6000kg (13,300lb) of fuel in its internal fuel tanks (shown in orange).

Twin Pratt and Whitney F100-PW-100 engines.

This smaller tank is called the trailing edge fuel tank.

The largest tanks are the ones in the middle of the wing.

Leading edge fuel tank.

Central fuselage fuel tanks.

This pipe can be used to jettison, or get rid of, fuel in an emergency.

Fuel can be pumped from one tank to another to help maintain the aircraft's trim, or balance.

Wind Tunnel

As scientists and engineers have learned more and more about aerodynamics, they have been able to build planes that fly faster, farther and more safely. Wind tunnels are used to study aerodynamics for both planes and cars (see page 17).

Even the very earliest planes were first tested as scale models inside simple wind tunnels. Wind tunnels today are much more complicated but the principles are still the same. Air is blown over a scale model of the aircraft, or part of it, and engineers see the effect this has. Computers are used to monitor results.

This airliner model is being tested in a high speed wind tunnel.

Adding fuel in-flight

The plane's fuel tank is usually filled on the ground where tanker trucks pump fuel into the tanks at incredibly fast speeds.

Some military planes, though, can take on extra fuel while flying. A large tanker aircraft, often a converted airliner or cargo plane, flies close to the plane in need of fuel. This requires great skill from both pilots.

This VC10 tanker can supply planes with up to 86,000kg (190,000lb) of fuel through its three fuel probes.

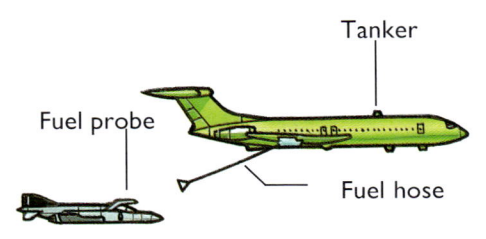

Tanker

Fuel probe

Fuel hose

The two aircraft fly extremely close to each other. The tanker extends its hose, the other plane extends its fuel probe.

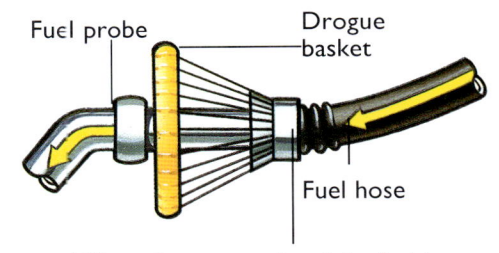

Fuel probe

Drogue basket

Fuel hose

When the two probes join, fuel is pumped quickly from the tanker to the plane that needs fuel.

47

Learning to fly

Becoming a pilot takes a long time and lots of hard work. This is because pilots have to know so much information about the aircraft they fly and about other important things, such as weather, maintainance, safety and navigation. Trainee pilots are tested on many lessons in the classroom as well as flying with an instructor in an aircraft called a trainer.

These trainees at a flying school are being taught about radio communications.

Aircrew, as well as pilots, need training. Here, aircrew learn what to do if there's a fire.

Trainers

Trainers often have two sets of controls so that the instructor can take over if the trainee pilot is having difficulty. There are trainers for all different types of aircraft.

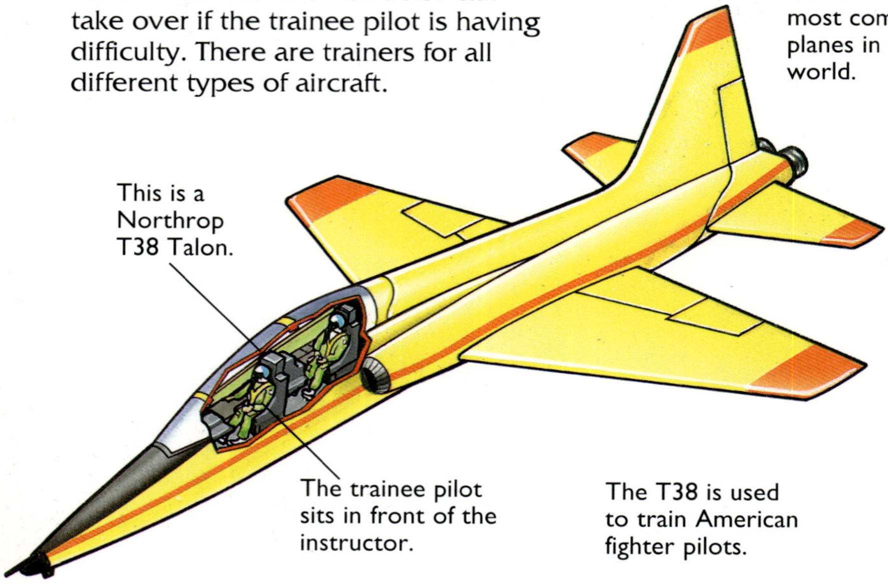

This is a Northrop T38 Talon.

The trainee pilot sits in front of the instructor.

The T38 is used to train American fighter pilots.

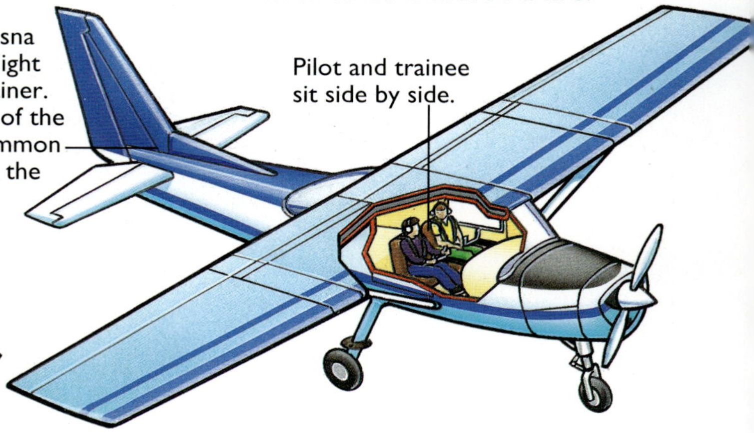

This Cessna 150 is a light plane trainer. It is one of the most common planes in the world.

Pilot and trainee sit side by side.

Going solo

After many hours in the classroom and up in the air with an instructor, a trainee pilot must fly the aircraft alone, without the instructor's help. This is called going solo. Once trainees have flown solo for a certain time, have learned about flying cross-country and passed various tests, they will be given their first pilot's licence.

Gliding

Some people learn to fly in a type of plane without an engine called a glider or sailplane. Learning to glide is cheaper and easier than learning to fly a powered aircraft but it is still extremely exciting and demanding.

Gliders can be launched into the air in many ways. A few gliders have small engines just for take-off. Many gliders are launched using a powerful winch on the ground. The winch pulls the glider along fast enough for the air rushing over the wings to provide the lift needed for take-off.

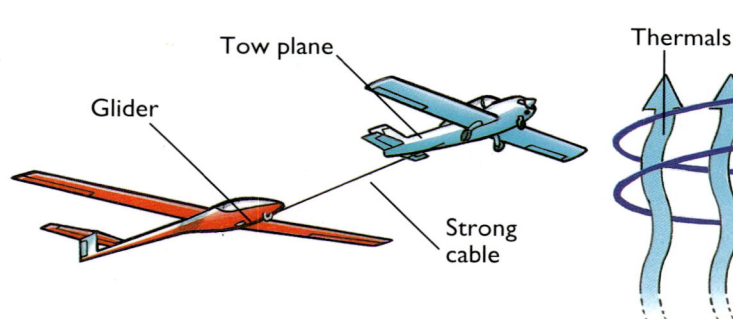

Tow plane

Glider

Strong cable

Thermals

Some gliders are towed into the air by a light aircraft known as a tug or tow plane. At an agreed height, the glider is released and the tow plane flies back to the airfield.

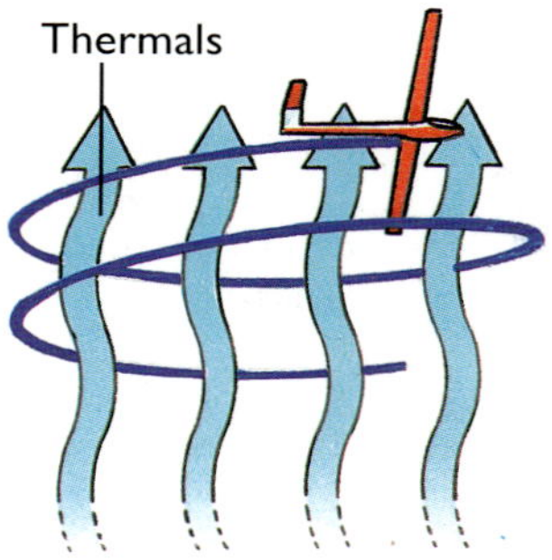

Once in the air, gliders circle around on pockets of warm air called thermals. An experienced pilot in good conditions can keep a glider flying for many hours.

Gliders are built from strong but very light materials to keep their weight down. These long wings provide lots of lift.

Single wheel landing gear

This glider's cockpit seats two people.

T-shaped tail with the elevators on top of the tail fin.

Aileron

Flight simulators

Flight simulators are complex machines which make you feel you are really flying. They do this by using a mixture of projector displays, realistic movement, and a complete copy of the real plane's cockpit. Flight simulators are not just flown by trainees. Experienced pilots also use them when learning to fly new aircraft or when working on how to cope with an emergency, which would be difficult or dangerous to do in a real aircraft.

This simulator is used by Japanese airline, JAL, to train their pilots.

Here is an example of the sort of realistic scene which can be generated by computers and viewed from inside an advanced simulator.

There are many different types of flight simulator. The most advanced, like this model at the National Aerospace Laboratory in Holland, are controlled by computers and *hydraulics**.

This part is called the dome. Computers and projectors display realistic pictures onto the inside surface of the dome.

Computers which control the simulator (see page 7).

Different cockpits can be put inside the dome. This cockpit is from an American F16 jet fighter.

The legs are powered by these strong hydraulic systems which react quickly to the pilot's controls.

This is one of the legs which tilt the simulator in all directions. They give the impression of turning, climbing and descending.

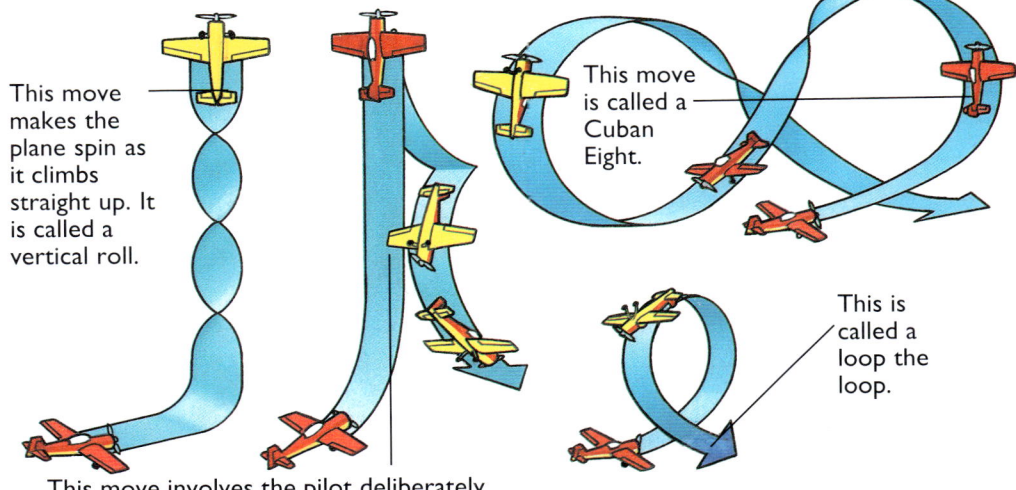

Details about how the pilot flies can be recorded and looked at later.

Platform

Aerobatics

Tricks and special moves performed by a pilot in a plane are called aerobatics. It takes a long time for even experienced pilots to learn how to do them.

This Pitts Special is a very rugged biplane often used for aerobatics.

This move makes the plane spin as it climbs straight up. It is called a vertical roll.

This move is called a Cuban Eight.

This is called a loop the loop.

This move involves the pilot deliberately *stalling** the plane so that it drops out of the sky. It is called a stall turn.

49

Take-off and landing

Nearly all planes need a long stretch of flat ground or calm water for take-off and landing. Take-off and landing are a pilot's busiest times.

How a plane takes off

Taking off needs more power than any other part of a flight. A plane at take-off is at its heaviest because it hasn't used much of its fuel. The engines must move the plane fast enough for the wings to overcome the plane's weight and lift the plane into the sky. Below is a diagram of how a plane takes off and lands.

Floats and skids

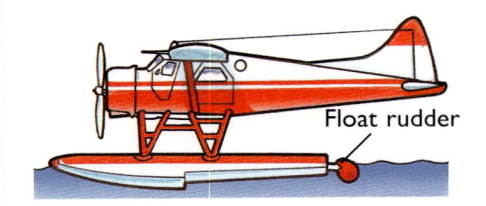

This Canadian DHC-2 Beaver can be fitted with floats or skids.

Some aircraft don't need a runway for take-off and landing. Their landing gear wheels (see below) may be replaced with skids for landing on ground; or replaced with floats, underneath their fuselage or wing, for landing on water. These floats usually have small rudders on the back to allow the pilot to steer the plane once it is on the water.

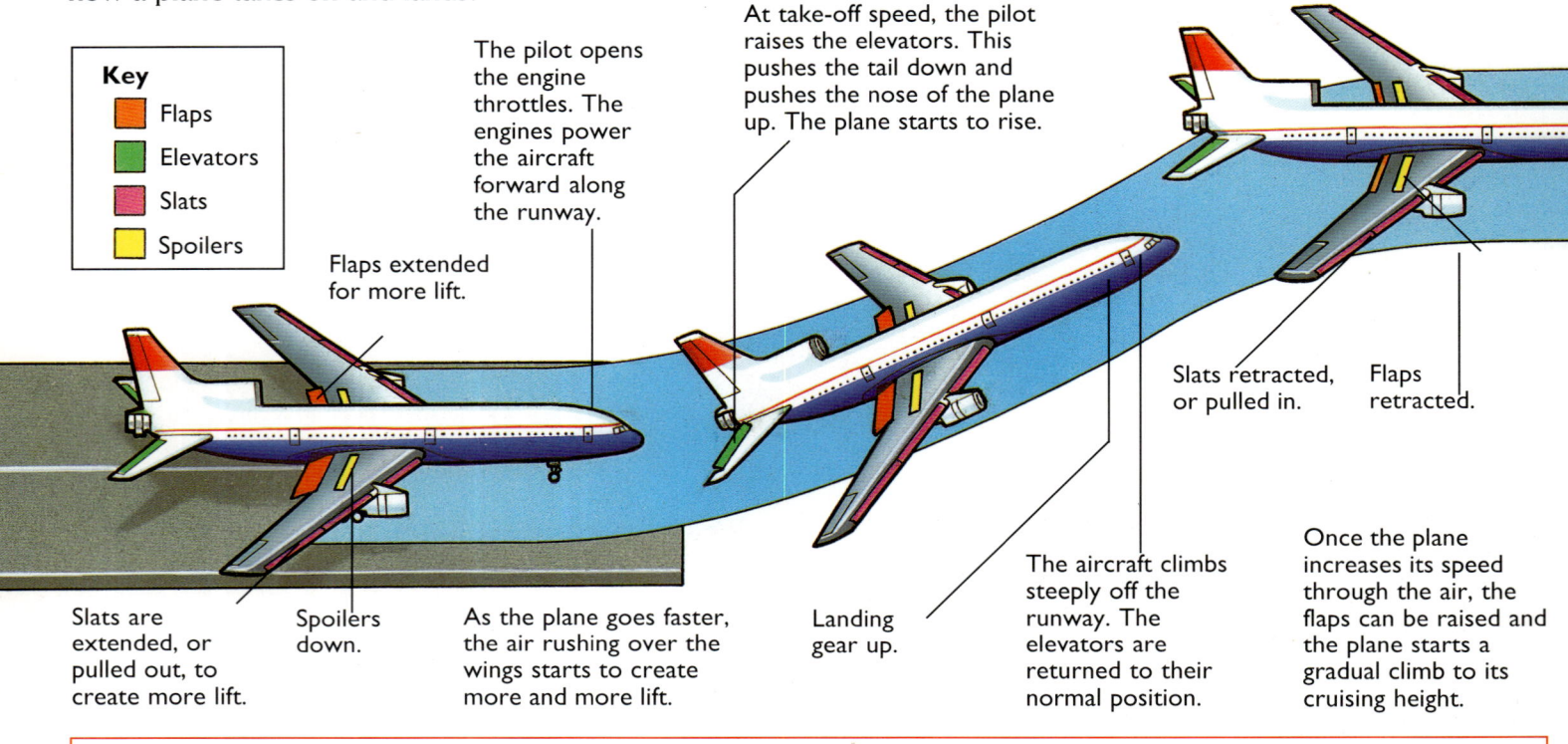

Key
- Flaps
- Elevators
- Slats
- Spoilers

The pilot opens the engine throttles. The engines power the aircraft forward along the runway.

Flaps extended for more lift.

Slats are extended, or pulled out, to create more lift.

Spoilers down.

As the plane goes faster, the air rushing over the wings starts to create more and more lift.

At take-off speed, the pilot raises the elevators. This pushes the tail down and pushes the nose of the plane up. The plane starts to rise.

Landing gear up.

The aircraft climbs steeply off the runway. The elevators are returned to their normal position.

Slats retracted, or pulled in.

Flaps retracted.

Once the plane increases its speed through the air, the flaps can be raised and the plane starts a gradual climb to its cruising height.

The landing gear

The parts of the plane that touch the ground at take-off and landing, and the parts that support them, are called the landing gear. As a plane lands, the landing gear is put under great strain. Modern landing gears are very strong. For example, the entire Boeing 747 landing gear is tested to support almost double the plane's actual weight, which is an incredible 360,000kg (800,000lb).

This is part of the landing gear of a Falcon 900 business jet.

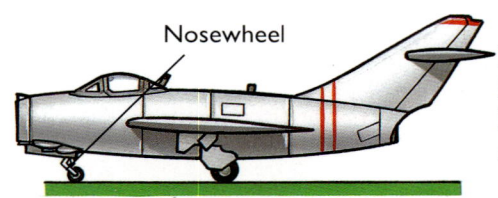

Nosewheel

The most common form of landing gear has three sets of wheels and is called a tricycle landing gear. Both these planes have this sort of landing gear.

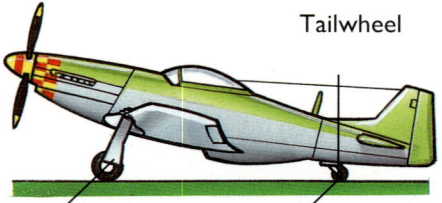

Tailwheel

Two sets, the main gear, are placed under the body or wings of the plane.

The third set is either under the nose or tail to keep the plane level.

Retractable landing gear first became common in the 1930s. It helps make the plane much more *streamlined**.

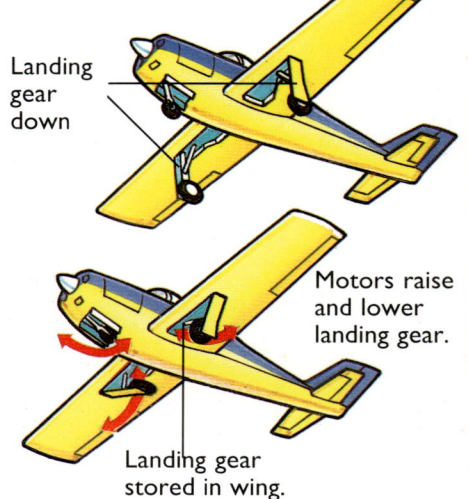

Landing gear down

Motors raise and lower landing gear.

Landing gear stored in wing.

How a plane lands

The pilot carefully sets a course to fly down to the runway. As the plane starts to descend, the pilot makes small changes to the position of the elevators and the power of the engines. When the plane is within 5 or 6m (16-20ft) of the ground, the pilot reduces engine power and raises the elevators a little more. The plane gently hits the runway.

Take-off and landing at sea

On many aircraft carriers, a strong catapult driven by steam from the ship's boilers can pull an aircraft from a standstill to 240km/h (150mph) in under two seconds.

An arrester hook is fitted to many aircraft on ships. As the plane lands, the hook trails along the deck and catches on a strong set of cables which act as an anchor.

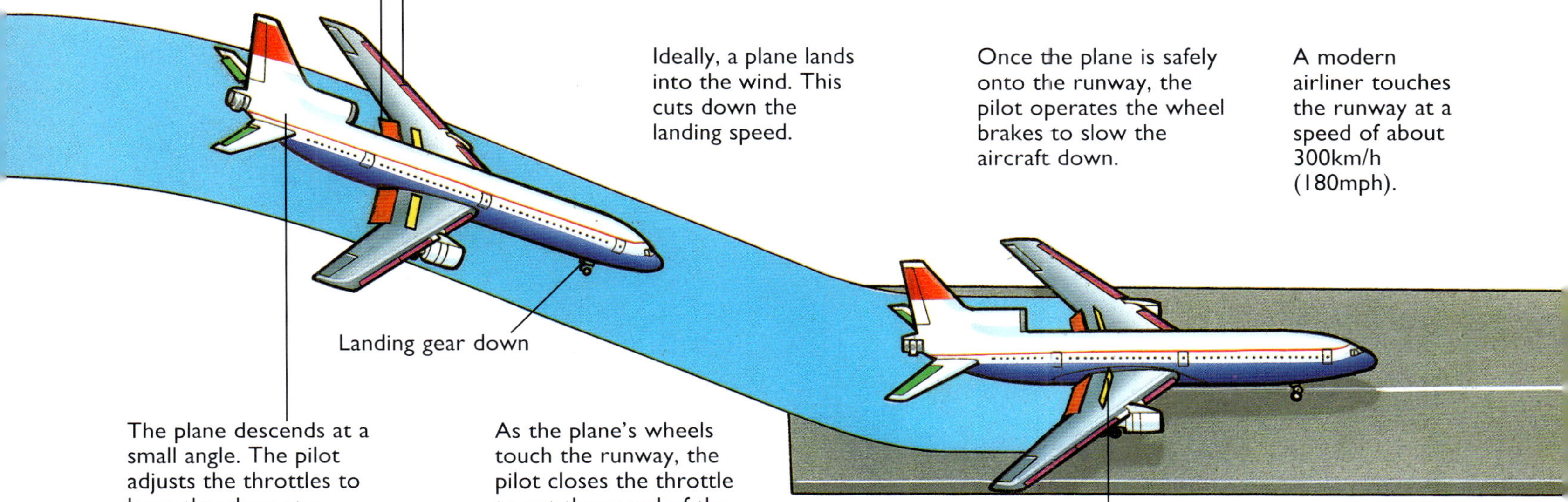

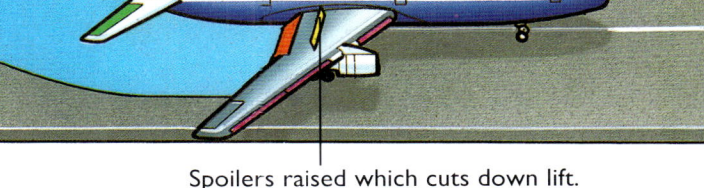

Slats and flaps out for extra lift.

Landing gear down

The plane descends at a small angle. The pilot adjusts the throttles to keep the plane at a steady speed.

As the plane's wheels touch the runway, the pilot closes the throttle to cut the speed of the engines right down.

Ideally, a plane lands into the wind. This cuts down the landing speed.

Once the plane is safely onto the runway, the pilot operates the wheel brakes to slow the aircraft down.

A modern airliner touches the runway at a speed of about 300km/h (180mph).

Spoilers raised which cuts down lift.

Stopping after landing

Modern aircraft land at high speeds. They have other methods, apart from the wheel brakes, to help them stop quickly.

Air brakes or spoilers can be on a planes's body or its wings (see page 37). They are flaps that open out to disrupt the flow of air. This creates plenty of drag which slows the aircraft down.

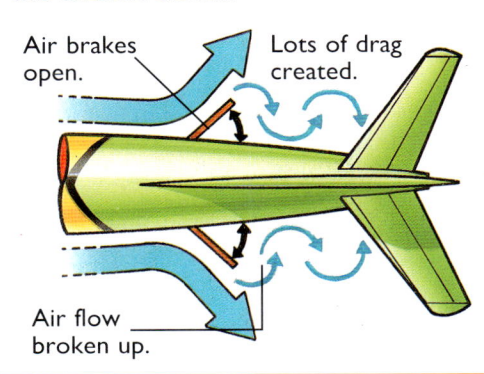

Air brakes open.
Lots of drag created.
Air flow broken up.

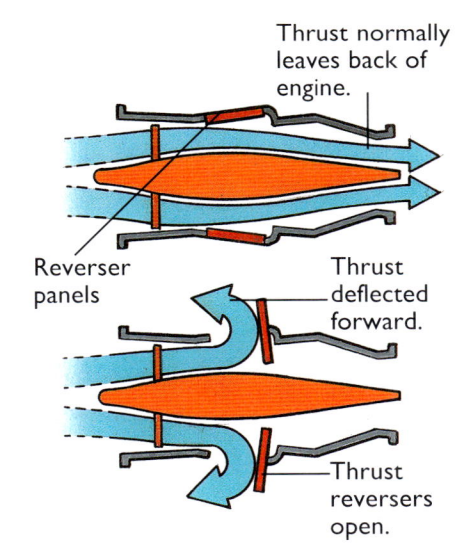

Thrust normally leaves back of engine.

Reverser panels

Thrust deflected forward.

Thrust reversers open.

Thrust reversers simply change the direction the engine's power is pushing. Instead of the power pushing the plane forward, it pushes against the way the plane is moving, so the plane slows down.

Some planes are fitted with brake parachutes. These open up from the back of the aircraft and create a lot of drag which helps slow the plane down. A brake parachute on a military jet fighter may take only two seconds to inflate fully but can cut 25% off the braking distance in dry conditions and as much as 50% in the wet.

This F117A stealth fighter has used its brake parachute on landing.

VSTOL planes

VSTOL stands for Vertical or Short Take-Off and Landing. A VSTOL plane can either use a short stretch of runway or can take off and land straight up and down from a small space such as a clearing in a forest or jungle. Much of the time, VSTOL planes use a runway for a short take-off but land vertically. This saves a lot of fuel.

This machine, called the Flying Bedstead, was built by Rolls Royce to test engines for VSTOL. It was first used in 1953.

BAe Sea Harrier Mk1

The Harrier is the best known VSTOL plane. It was first built by British company Hawker Siddeley (later British Aerospace) and Harriers have served in the British, Indian, Italian, Spanish and United States military services. The Sea Harrier is a version of the Harrier used by navies. It can take off and land from the decks of aircraft carriers and has a top speed of 1,120km/h (720mph).

This Sea Harrier is landing vertically onto the deck of an aircraft carrier.

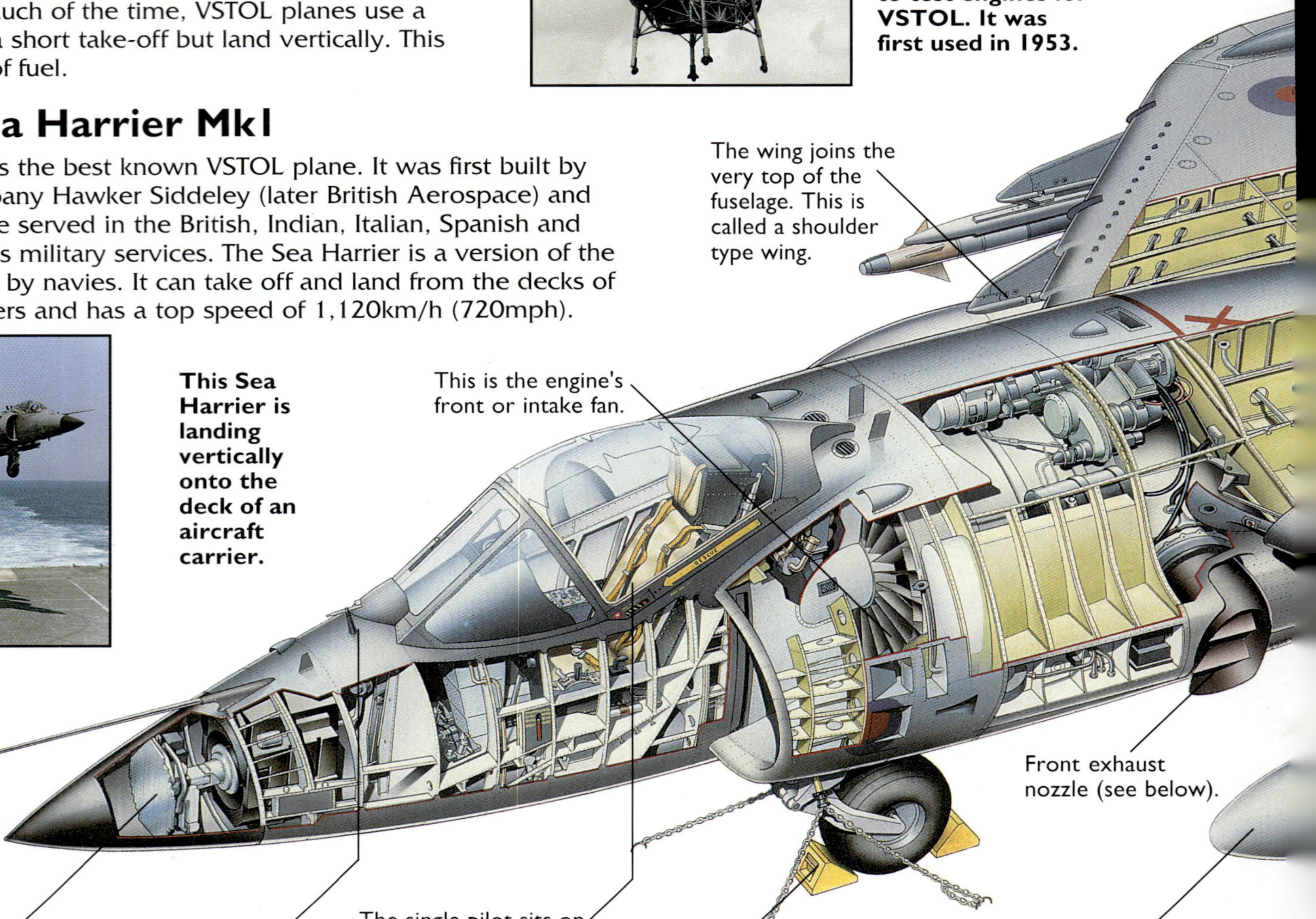

The wing joins the very top of the fuselage. This is called a shoulder type wing.

This is the engine's front or intake fan.

Front exhaust nozzle (see below).

Advanced Blue Fox radar is stored in the nose.

This measures how much the plane yaws (see page 36). It is called a yaw vane.

The single pilot sits on a Martin Baker Type 10 ejection seat. He can eject safely even from the ship's deck.

Wedges called chocks stop the plane from moving when parked on the ground.

This external fuel tank holds more fuel than the tanks of 12 family cars.

Pegasus engine

The Harrier is powered by a single turbofan engine called the Rolls Royce Pegasus. The main difference between the Pegasus and other turbofans is that the direction the gases leave the engine can be controlled. This is called vectored thrust (see opposite page). The engine has four linked exhaust nozzles through which the hot gases and cold air are forced out. These nozzles can be turned around by the pilot from the cockpit.

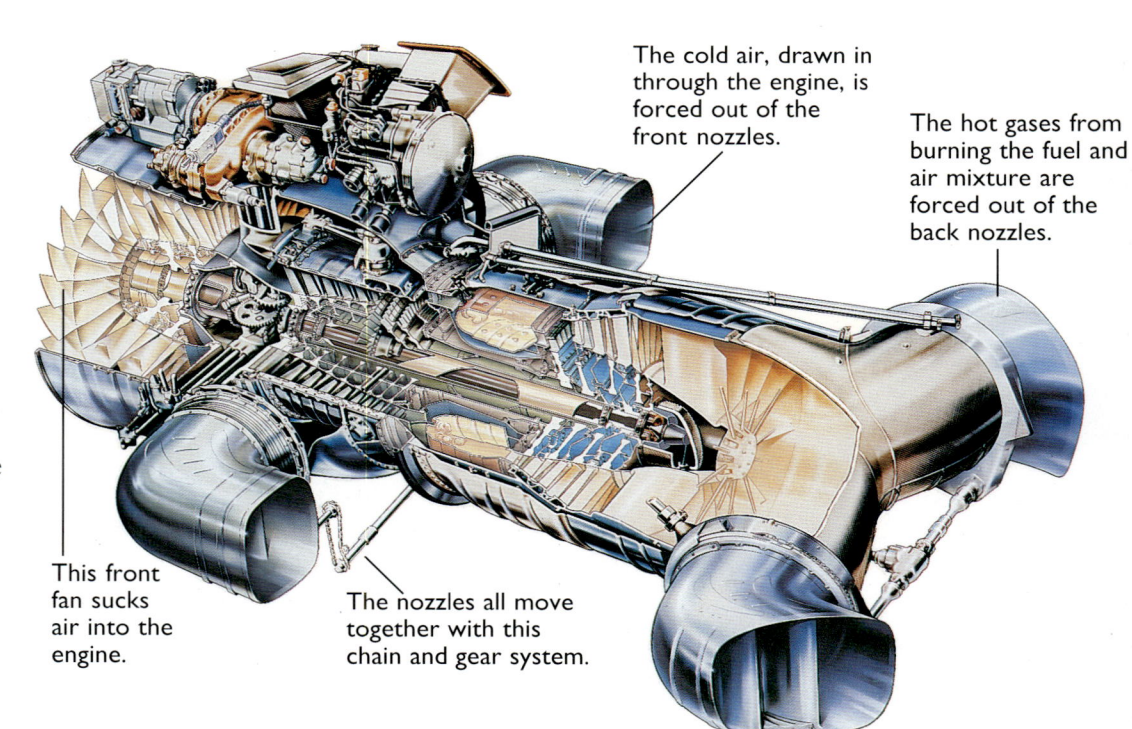

This front fan sucks air into the engine.

The nozzles all move together with this chain and gear system.

The cold air, drawn in through the engine, is forced out of the front nozzles.

The hot gases from burning the fuel and air mixture are forced out of the back nozzles.

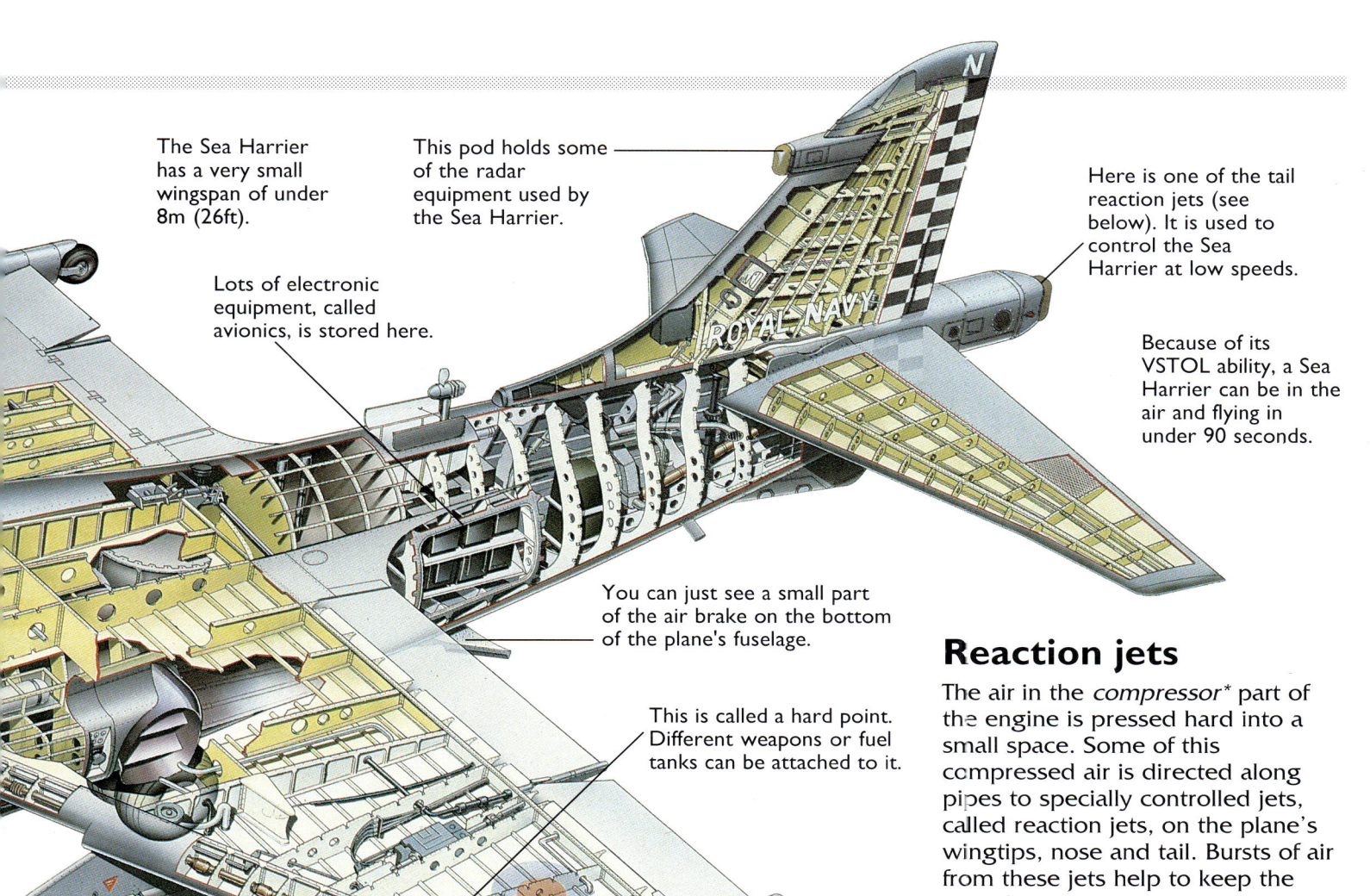

The Sea Harrier has a very small wingspan of under 8m (26ft).

This pod holds some of the radar equipment used by the Sea Harrier.

Here is one of the tail reaction jets (see below). It is used to control the Sea Harrier at low speeds.

Lots of electronic equipment, called avionics, is stored here.

You can just see a small part of the air brake on the bottom of the plane's fuselage.

Because of its VSTOL ability, a Sea Harrier can be in the air and flying in under 90 seconds.

This is called a hard point. Different weapons or fuel tanks can be attached to it.

This wingtip wheel helps steady the plane when it is on the ground.

Reaction jets

The air in the *compressor* part of the engine is pressed hard into a small space. Some of this compressed air is directed along pipes to specially controlled jets, called reaction jets, on the plane's wingtips, nose and tail. Bursts of air from these jets help to keep the aircraft stable when it is hovering or when it is moving slowly.

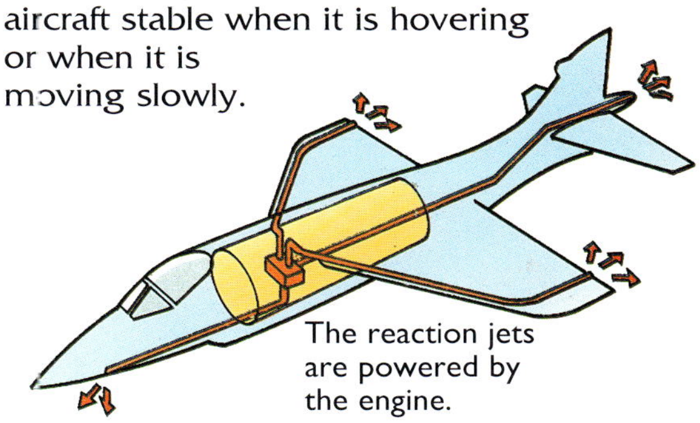

The reaction jets are powered by the engine.

Vectored thrust

Vectored thrust works by nozzles directing the flow of gases out of the engine. The engine can produce up to 88kN (19,800lb) of thrust. Most or all of this is needed for vertical take-off. If the plane weighs more than the thrust can lift straight up, it cannot take off vertically and must use a ramp or runway instead.

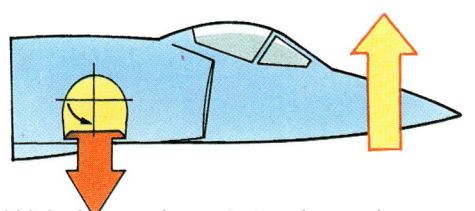

With the nozzles pointing down, the gases are forced down which provides up-thrust, lifting the plane off the ground.

When the nozzles are set pointing back, the gases are forced back which powers the aircraft forward.

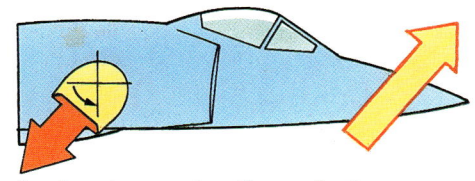

Angling the nozzles diagonally down moves the aircraft both up and forward at the same time.

VIFF'ing

Paddles around the engine's exhaust angle the exhaust's direction.

Pilots can use vectored thrust while flying. By altering the direction of the exhaust, a plane can make tighter turns and climb faster than normal. This is called VIFF'ing (Vectoring In Forward Flight). Some military fighters which are not VSTOL aircraft are now being fitted with equipment which allows them to VIFF. This will make them faster and easier to move around in combat.

Safety

Early planes were often dangerous to fly. They were difficult to control and had few safety features or instruments to help the pilot. Modern aircraft are not only easier to control but also contain many instruments designed to avoid accidents. For example, radar warns of aircraft or other problems ahead.

The flight data recorder, often called the black box, records all the details of a flight.

The black box is examined by experts if there's a crash or other problem.

Passenger safety

Modern airliners are designed carefully to avoid accidents and can still fly even if one engine stops working. They all have many features built in to help passengers survive and escape from the plane if it crashes.

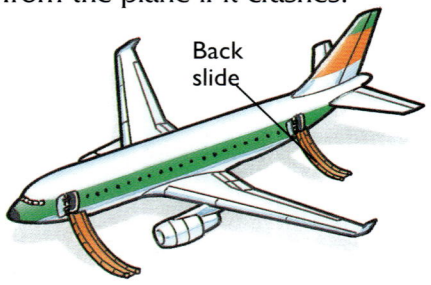

Back slide

These slides are a quick and safe way to leave the aircraft. They are filled with air and are worked by the airliner's aircrew.

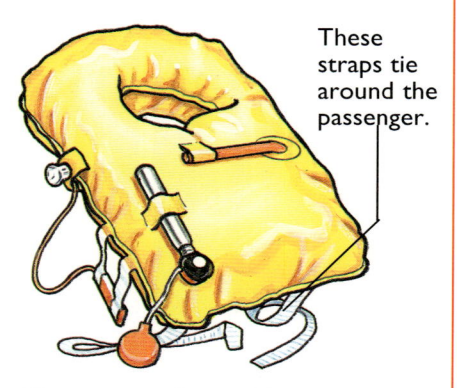

These straps tie around the passenger.

All passengers have a life jacket by or under their seats. They inflate automatically when you pull a tab.

Life rafts provide safety and shelter for survivors if a plane crashes in water. This raft, made by RFD Equipment, holds up to 9 people.

Ejection seats

Jet aircraft fly too fast for pilots and crew members to jump out of the plane and use a parachute. So an ejection seat is used to fire crew members safely out of the plane and away from danger.

Since they were introduced in the 1940s, ejection seats have saved over 10,000 lives. Ejection seats made by one British company alone, called Martin-Baker, have saved over 7000 people.

This Mk16 ejection seat was built by Martin Baker. It is used in both the Eurofighter and the Dassault Rafale military jets. Although it weighs less than a man, the ejection seat can safely eject the pilot from any altitude including ground level. The whole ejection process is incredibly fast. In 0.25 seconds the pilot is out of the aircraft and in under 3 seconds, the main parachute is fully opened.

Parachute container

The straps in the harness are tightened automatically.

Firing handle

Guide rails

Oxygen supply

Leg straps tighten to help secure pilot firmly in seat.

Rocket pack

How ejection seats work

The pilot pulls the firing handle. This sets off a computer-controlled sequence. The seat straps tighten automatically, and all or part of the plane's canopy is fired away.

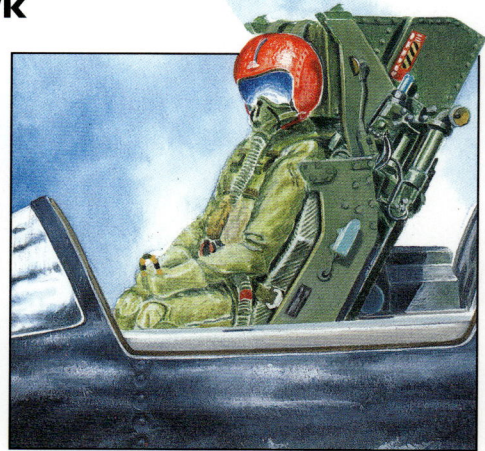

The pilot's radio and oxygen are disconnected and replaced by emergency ones built into the seat. The ejection gun fires the seat up the guide rails and free of the plane.

Pilot safety

Pilots and aircrew flying military jets need to be well-protected. They are often flying at very high altitudes and need oxygen to breathe. They usually sit on a pack of equipment that contains tools, a life raft, and even food to help them survive if they crash in an isolated place.

When a fast jet makes a tight turn or pulls out of a steep dive, the force of *gravity** presses hard on the pilot. An anti-G suit stops the force from hurting him. It is filled with air and is worn over the flying suit.

The knee parts of the G-suit are cut out to allow the pilot to bend his knees more easily.

This is the oxygen face mask. The ejection seat has 15-20 minutes supply of oxygen on board.

The tinted sun visor protects the pilot from strong glare.

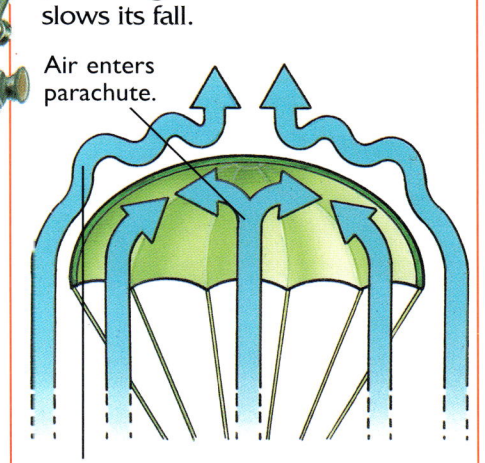

Oxygen supply pipe

These leather gloves are fire-resistant.

The pilot's flying suit and underwear are made from material that is hard to set fire to.

When just clear of the aircraft, a rocket pack under the seat fires for about a fifth of a second. This pushes the seat about 100m (330ft) away from the plane.

An explosive charge opens a small parachute called a drogue. The drogue parachute helps slow the seat down and, on some ejection seats, pulls out the main parachute.

Once the main parachute is deployed, or fully opened, the seat falls away from the pilot who is then left to make an ordinary safe parachute landing.

Parachutes

Parachutes slow down a person falling through the air. This means that he or she can land safely when bailing (jumping) out of a plane. A parachute is a large canopy made of flame-resistant material. When open, it creates lots of drag as air pushes against its underside. This slows its fall.

Air enters parachute.

Drag created by catching the air.

Parachutes are tightly fitted into packs either worn by the aircrew or stored on the ejection seat. Sometimes, a second, smaller reserve pack is attached just in case the main one doesn't work.

Parachutes are used for other jobs too. They can slow fast planes down when they land (see page 51). Parachutes are also put on supplies and food so they can be dropped to places without runways or roads.

Modern military planes

Modern military planes come in all shapes and sizes and perform lots of different jobs. Many of them are designed either to defend an area against an enemy attack or strike at a target with weapons. Not all military aircraft are used just for combat though. Some are used for training, rescue work and transport. Others are used to gather information about an area. This is called reconnaissance. Military reconnaissance planes photograph and check on enemy sites and troop movements.

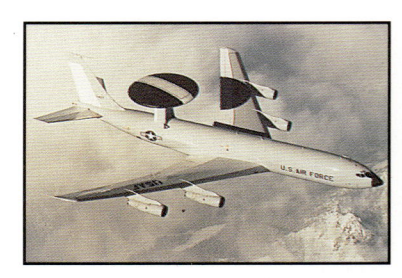

This C17 Globemaster transports troops and tanks.

This Boeing AWACS is used for reconnaissance.

Panavia Tornado GR.1

This military jet was built by a group of companies from Germany, Italy and Great Britain. It is used for a number of different tasks and is called a Multi-Role Combat Aircraft (MRCA). This GR.1 version is used mainly as a strike aircraft but also sometimes for reconnaissance.

Here is a GR.1 used by the Royal Air Force in Britain.

This probe collects information about the air and sends it to one of the plane's computers.

The pilot sits at the front. Behind him, the second seat is for the navigator who also operates the weapons.

Cockpit canopy

This is an air intake. You can see how it is made up of a frame of metal strips all joined together.

This radar antenna maps the ground as the Tornado flies along.

The nosecone unhinges to allow engineers to reach the radar equipment stored there.

Inside this pod, advanced electronics and a laser find the distance to a target. It is called a laser rangefinder.

This is the fuel tank of one of the Tornado's missiles.

Swing wings

Straighter wings provide more lift and control which is important for take-off and landing. However, for fast speeds, wings swept back at an angle are best. One solution is to build wings which can swing from one position to another. These are called swing wings.

The Tornado's wings can be moved into one of four different positions, 67°, 58°, 45° and 25°.

67° 58° 45° 25°

The Tornado is a small plane but it weighs more than many bigger aircraft. This is partly because of the heavy load it carries. Almost half of the Tornado's weight is fuel and weapons.

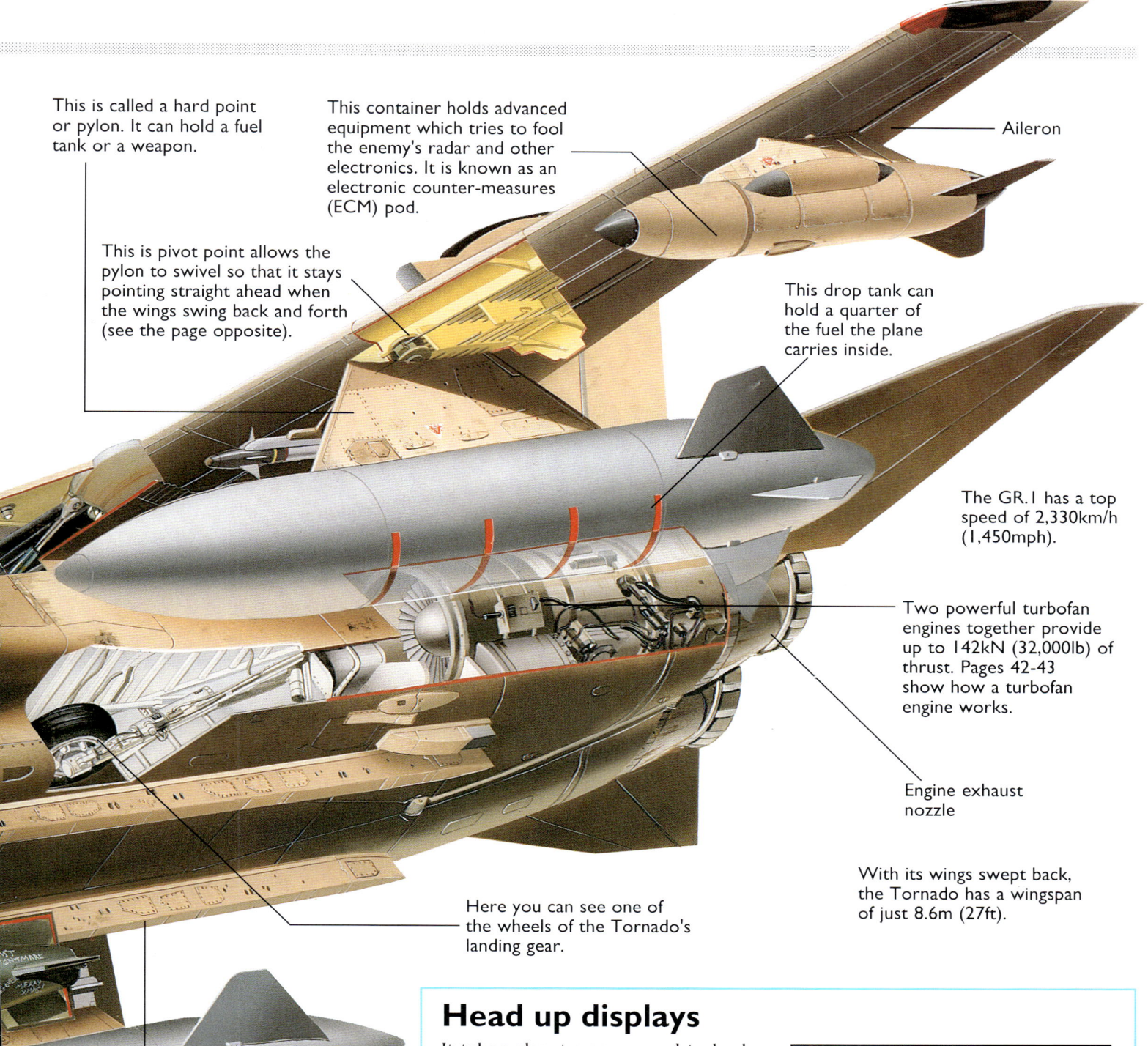

This is called a hard point or pylon. It can hold a fuel tank or a weapon.

This container holds advanced equipment which tries to fool the enemy's radar and other electronics. It is known as an electronic counter-measures (ECM) pod.

Aileron

This is pivot point allows the pylon to swivel so that it stays pointing straight ahead when the wings swing back and forth (see the page opposite).

This drop tank can hold a quarter of the fuel the plane carries inside.

The GR.1 has a top speed of 2,330km/h (1,450mph).

Two powerful turbofan engines together provide up to 142kN (32,000lb) of thrust. Pages 42-43 show how a turbofan engine works.

Engine exhaust nozzle

With its wings swept back, the Tornado has a wingspan of just 8.6m (27ft).

Here you can see one of the wheels of the Tornado's landing gear.

Missiles can be launched from these hard points attached to the bottom of the fuselage.

From take-off, the Tornado takes under two minutes to climb to over 9,000m (30,000ft).

Head up displays

It takes about one second to look down at instruments before bringing your head up again to look ahead. This may be fine in a slow vehicle but in a jet flying at 2,000km/h (1,250mph) it could be extremely dangerous. In that time, the plane could have flown over 500m (530 yards).

Head Up Displays, or HUDs for short, are found in many of today's modern military fast jets. They allow information to be displayed straight in front of the pilot either on a clear screen in the cockpit or directly onto the pilot's visor. The screen is see-through so that the pilot can look past the details at the view outside without moving his head.

The projector in this helmet displays important information onto the pilot's helmet visor.

57

Navigation

The pilots of early planes had little more than maps and a compass to find their way with. They had to plan their course by watching the ground for landmarks. This is called dead reckoning. Today, pilots occasionally use dead reckoning but they usually rely on radio and radar navigation, map computers, and instruments which show the plane's height, speed and heading.

This is a Magellan EG-10 portable electronic map. It can be used in small civil aircraft which don't have expensive map computers built into their cockpit.

Gyroscopes

There are two main devices which help detect a change in the direction of the plane's movement. These are gyroscopes and accelerometers.

A gyroscope used on an aircraft is a fast-spinning wheel joined to a frame. The wheel always stays in exactly the same position, while the frame moves with the plane. The direction and distance that the frame moves is turned into a reading, telling the pilot the change in the aircraft's movement.

Gyroscopes are used in important navigational instruments like the artificial horizon and the directional gyro, which is a type of non-magnetic compass.

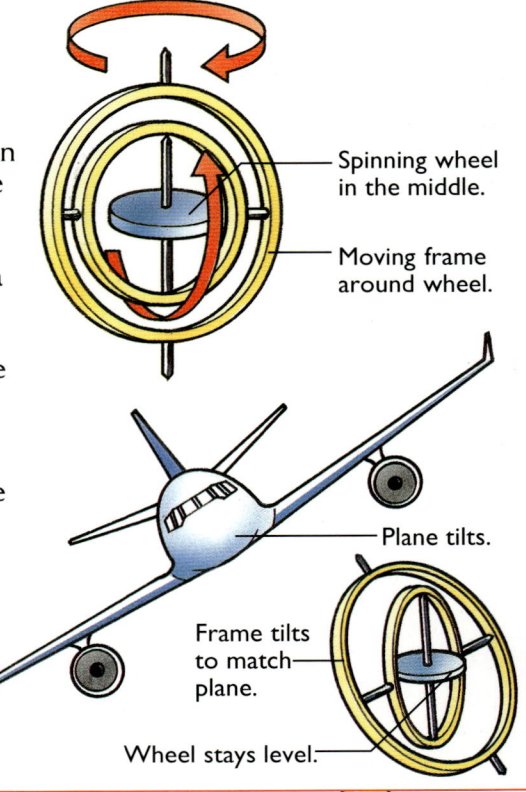

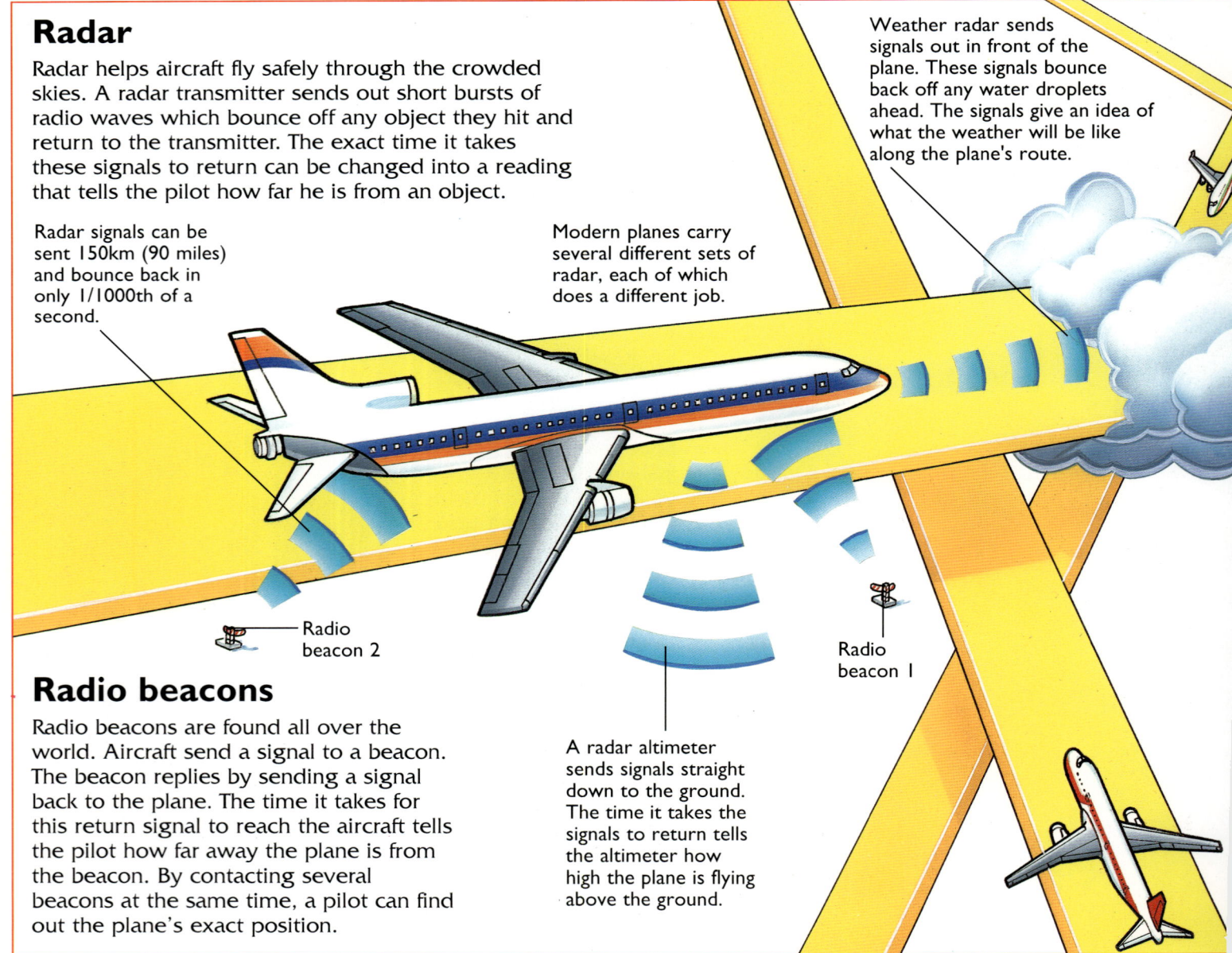

- Spinning wheel in the middle.
- Moving frame around wheel.
- Plane tilts.
- Frame tilts to match plane.
- Wheel stays level.

Radar

Radar helps aircraft fly safely through the crowded skies. A radar transmitter sends out short bursts of radio waves which bounce off any object they hit and return to the transmitter. The exact time it takes these signals to return can be changed into a reading that tells the pilot how far he is from an object.

Radar signals can be sent 150km (90 miles) and bounce back in only 1/1000th of a second.

Modern planes carry several different sets of radar, each of which does a different job.

Weather radar sends signals out in front of the plane. These signals bounce back off any water droplets ahead. The signals give an idea of what the weather will be like along the plane's route.

Radio beacon 2

Radio beacon 1

A radar altimeter sends signals straight down to the ground. The time it takes the signals to return tells the altimeter how high the plane is flying above the ground.

Radio beacons

Radio beacons are found all over the world. Aircraft send a signal to a beacon. The beacon replies by sending a signal back to the plane. The time it takes for this return signal to reach the aircraft tells the pilot how far away the plane is from the beacon. By contacting several beacons at the same time, a pilot can find out the plane's exact position.

Accelerometers

An accelerometer is an electronic device which does a similar job to a gyroscope. One part of an accelerometer is in a fixed position and another part can move with the plane. Electricity produces a *magnetic field** between the two parts and any change in movement disturbs the field. This change is fed into a computer which calculates the amount of movement.

Magnetic fields created between the two parts of an accelerometer.

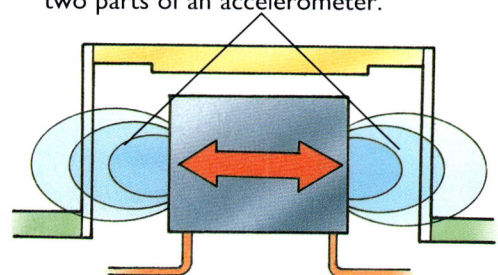

Autopilot

Most planes today have a system called an autopilot. It allows a plane to stay on a set course without the pilot having to hold the controls all of the time. An autopilot uses accelerometers and gyroscopes to detect movement. If the plane is moving off course, the autopilot's computer triggers electric or hydraulic motors, called servos.

These move the flight controls to bring the plane back on course. If the plane moves a large distance off course, the autopilot also warns the pilot. Autopilots are incredibly accurate, but there are some situations where the pilot's skill and experience are needed and autopilots cannot be used, such as flying through very bad weather.

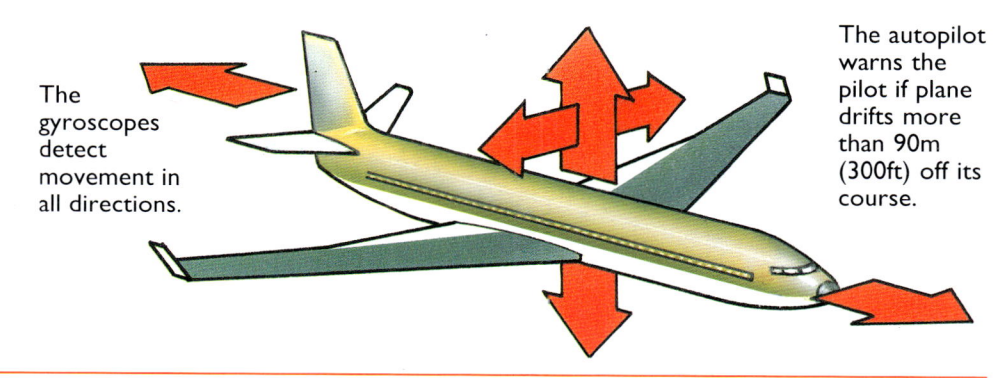

The gyroscopes detect movement in all directions.

The autopilot warns the pilot if plane drifts more than 90m (300ft) off its course.

Air corridors

With so many planes flying, important rules have been made to prevent collisions. Systems of aerial highways, called air corridors, have been created. Each plane flies along one specific corridor a safe distance away from any other planes.

Each corridor is about 15km (9 miles) wide and runs at a certain height above the ground.

There is usually a height difference of 500m (1600ft) between air corridors.

The descent towards the airport runway is called the approach.

Air traffic control

When flying, the pilot must obey the rules of the air, just as a car driver follows road traffic laws. Air traffic controllers manage aircraft in the skies and advise them on route changes or emergencies. Air traffic controllers use radios to speak to pilots but they also use two systems of radar. The first finds all the planes in the area a controller is scanning. The second receives an automatic signal from each aircraft which gives the controller the plane's identity, height and planned route. As a plane travels out of one area, it is handed from one controller to another.

This is part of air traffic control near London's Heathrow Airport.

Stacking and landing

When a plane gets close to a busy airport, it often has to join what is called a stack. This is like an aerial ladder made up of lots of different levels or rungs. When a plane arrives, it usually joins the stack at the highest point. The plane then travels around an oval-shaped route. Gradually, as aircraft below it land, the plane is ordered down the rungs of the ladder by the air traffic controllers. Eventually, it will reach the bottom level and then begin its approach to land.

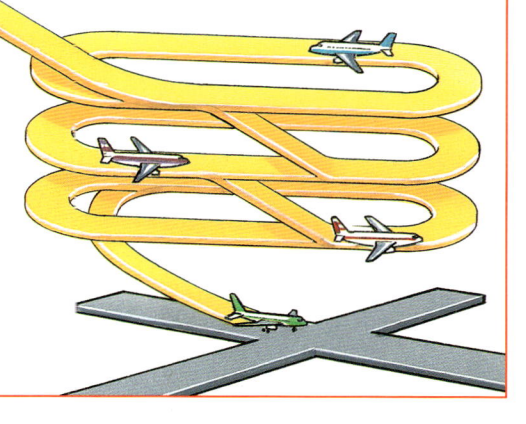

Stealth planes

Radar is very successful at spotting aircraft. Recently, engineers have found ways to make it harder for radar to see planes. Aircraft which are built with these anti-radar features are called stealth planes. Stealth confuses an enemy's electronics and radar equipment so the plane can fly by unnoticed. The United States has two stealth planes, the Northrop B2 bomber and the Lockheed F117A strike aircraft.

This is a F117A on a test flight over the United States.

Here you can see the Northrop B2's shape which is called a flying wing.

Lockheed F117A

This was the first stealth plane to be built and it first flew in 1981. It can travel secretly past heavily-defended enemy sites at night. The F117A is believed not to carry guns or air-to-air missiles. Instead, it relies on its stealth to escape harm.

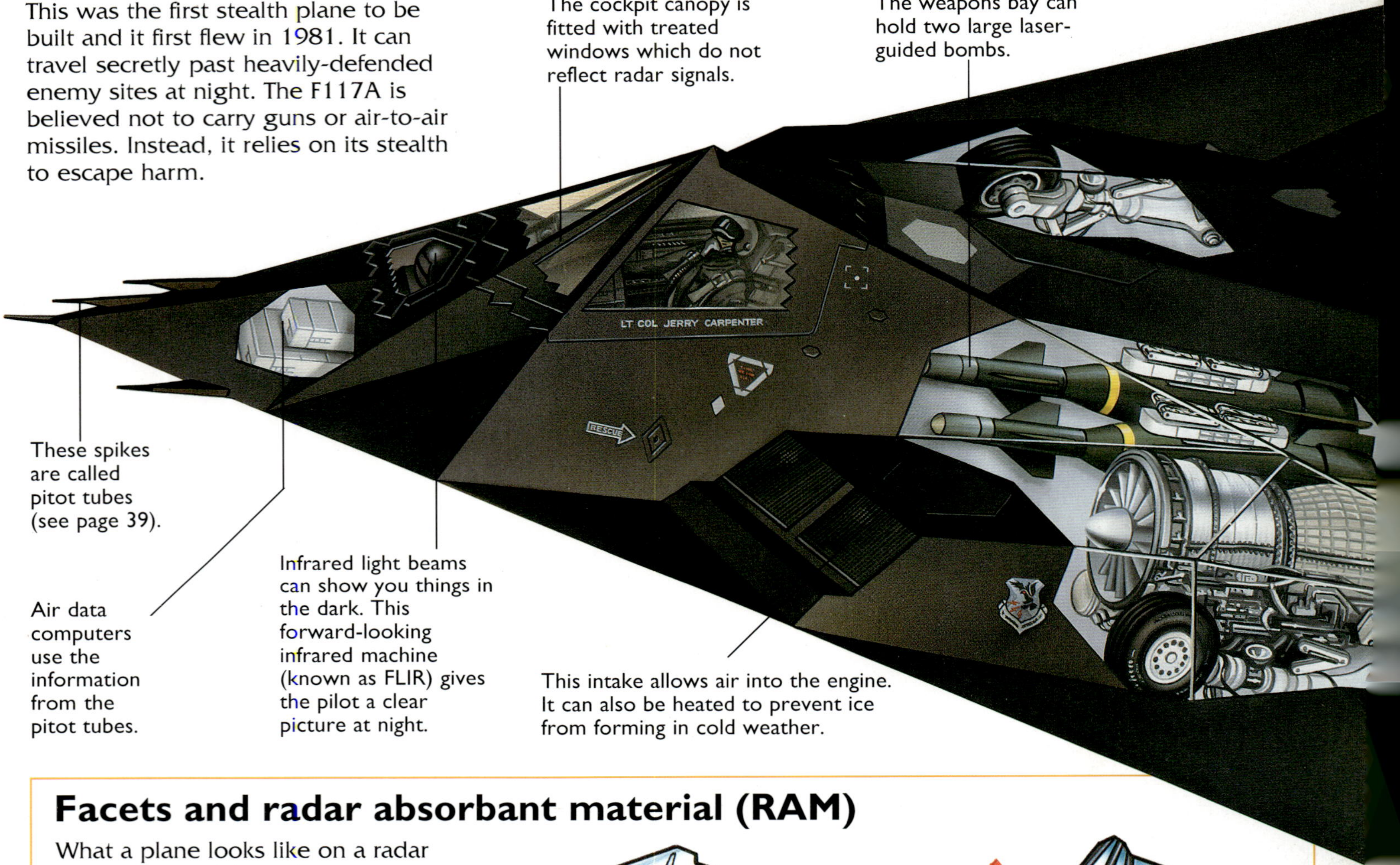

The cockpit canopy is fitted with treated windows which do not reflect radar signals.

The weapons bay can hold two large laser-guided bombs.

These spikes are called pitot tubes (see page 39).

Air data computers use the information from the pitot tubes.

Infrared light beams can show you things in the dark. This forward-looking infrared machine (known as FLIR) gives the pilot a clear picture at night.

This intake allows air into the engine. It can also be heated to prevent ice from forming in cold weather.

Facets and radar absorbant material (RAM)

What a plane looks like on a radar screen is called its radar cross-section, or RCS for short.

Radar bounces off large curved shapes best. So, instead, stealth planes are built with lots of differently-angled faces or facets. The facets are coated with a mixture of materials (called RAM) which absorb the radar signals rather than letting them bounce off. Together, facets and RAM make a stealth plane look more like birds than an aircraft on a radar screen.

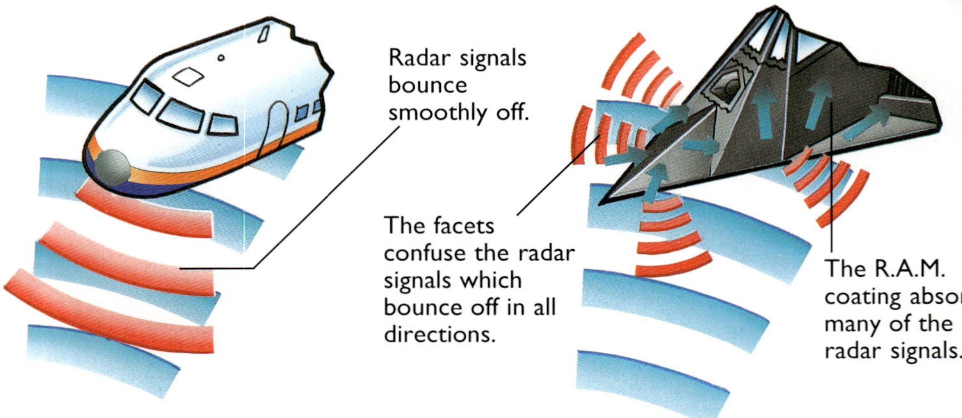

Radar signals bounce smoothly off.

The facets confuse the radar signals which bounce off in all directions.

The R.A.M. coating absorbs many of the radar signals.

On a normal aircraft, radar waves bounce off the large metal surfaces and back to the radar system.

On a stealth aircraft with facets, the signals bounce off at all angles and send back a confusing message to the radar station.

Avoiding the enemy

There are other ways to avoid being seen by the enemy apart from using stealth. Many warplanes since earliest times have been painted in shades and patterns to match the sky or ground, called camouflage. Some very fast jet planes use sheer speed to avoid being seen. Other planes, like the SR71 Blackbird, fly at extremely high altitudes.

The SR71 Blackbird is the fastest jet aircraft ever. Its top speed is over 3500km/h (2180mph).

This Sukhoi Su35 Russian jet fighter is camouflaged to match the winter landscape in Europe.

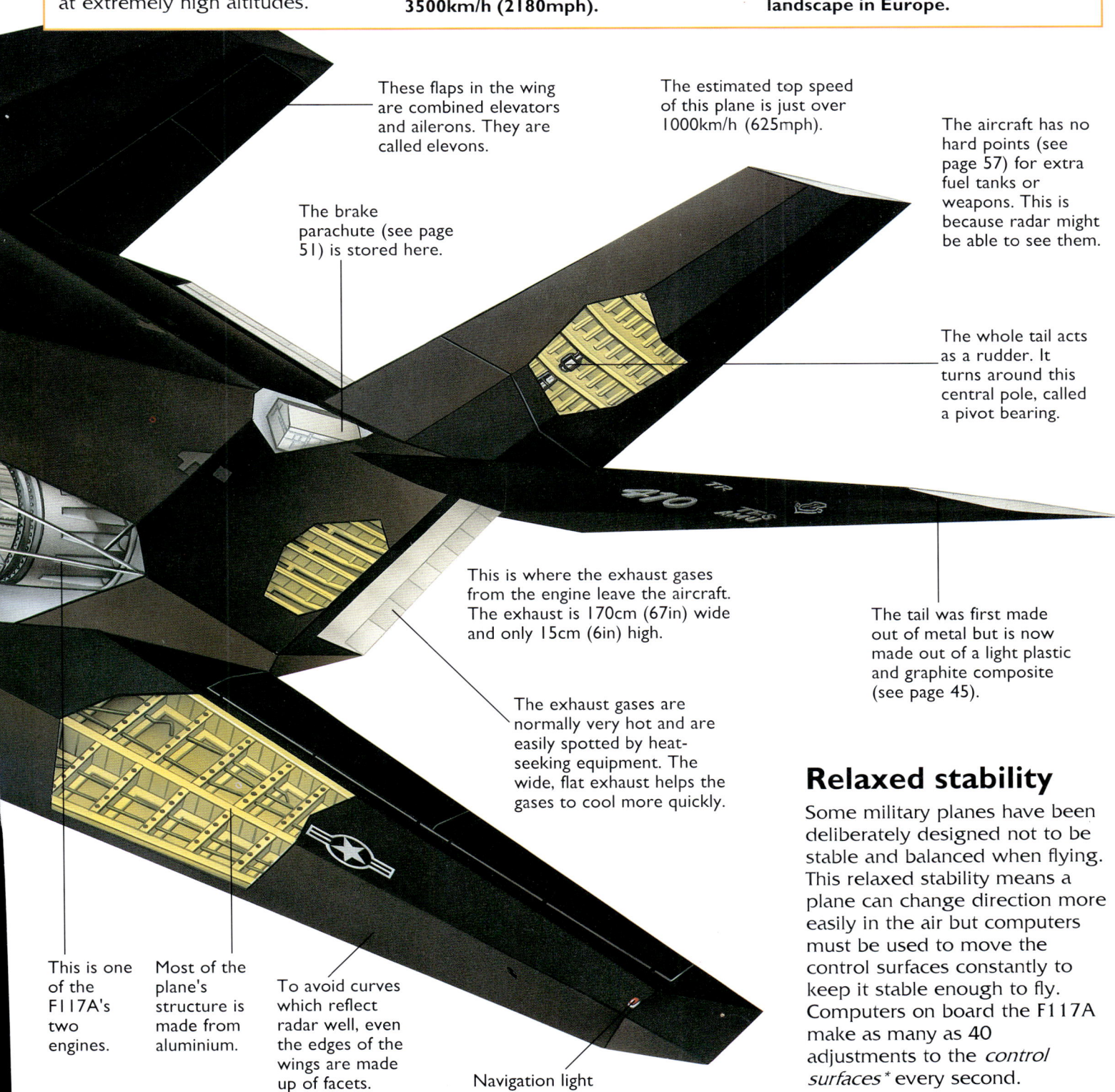

These flaps in the wing are combined elevators and ailerons. They are called elevons.

The brake parachute (see page 51) is stored here.

The estimated top speed of this plane is just over 1000km/h (625mph).

The aircraft has no hard points (see page 57) for extra fuel tanks or weapons. This is because radar might be able to see them.

The whole tail acts as a rudder. It turns around this central pole, called a pivot bearing.

This is where the exhaust gases from the engine leave the aircraft. The exhaust is 170cm (67in) wide and only 15cm (6in) high.

The tail was first made out of metal but is now made out of a light plastic and graphite composite (see page 45).

The exhaust gases are normally very hot and are easily spotted by heat-seeking equipment. The wide, flat exhaust helps the gases to cool more quickly.

This is one of the F117A's two engines.

Most of the plane's structure is made from aluminium.

To avoid curves which reflect radar well, even the edges of the wings are made up of facets.

Navigation light

Relaxed stability

Some military planes have been deliberately designed not to be stable and balanced when flying. This relaxed stability means a plane can change direction more easily in the air but computers must be used to move the control surfaces constantly to keep it stable enough to fly. Computers on board the F117A make as many as 40 adjustments to the *control surfaces** every second.

61

The future

In less than a century, aircraft have gone from being a dream to having a big impact on our lives. Nobody knows what will happen in another hundred years but it is possible to suggest what may happen in the near future.

Giant airliners

Aircraft makers will continue to look at ways of building airliners which are quieter, use less fuel and which can transport passengers at less cost.

This fuselage has two floors of seats.

This shape is called a clover-leaf body.

This design has two airliner fuselages joined together. It is called a horizontal double bubble.

More and more people want to travel by air and major airports cannot easily handle lots more flights. One solution is to build bigger airliners. Here are some possible airliner fuselage designs for the future.

Airbus A3XX

The Airbus A3XX is a giant airliner planned by Airbus Industries to hold as many as 830 passengers. It would not be much longer than a Boeing 747 but would have a much larger fuselage. There would be two floors for passengers and a bottom deck for cargo. As it will be made using some of the same parts as the Airbus airliners already in service, the A3XX could be flying by the year 2003.

A computer-generated picture of what the A3XX may look like.

Rocket engines

Rocket engines are used for space vehicles but a small number of aircraft such as the fastest ever plane, the Bell X-15, have been powered by rocket engines. Rocket engines are similar to gas turbine or jet engines (see pages 42-43) except that they don't have *turbines** and carry their own supply of oxygen on board.

The Bell X-15's top speed of 7300km/h (4530mph) makes it the fastest ever.

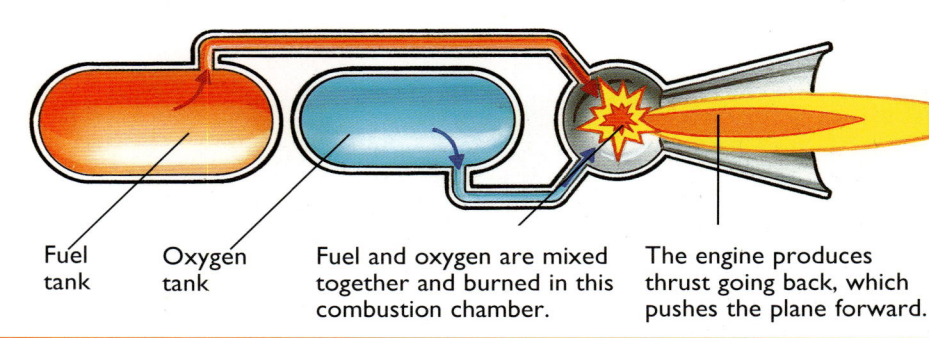

Fuel tank

Oxygen tank

Fuel and oxygen are mixed together and burned in this combustion chamber.

The engine produces thrust going back, which pushes the plane forward.

Space airliner

In the future, you may see airliners powered by a mixture of rocket and jet engines. They would be extremely fast and designed to fly partly in space and partly in the Earth's atmosphere. They would cut down the journey time from one side of the planet to the other to perhaps as little as two or three hours.

This is an artist's impression of a Japanese rocket airliner flying in the 21st century.

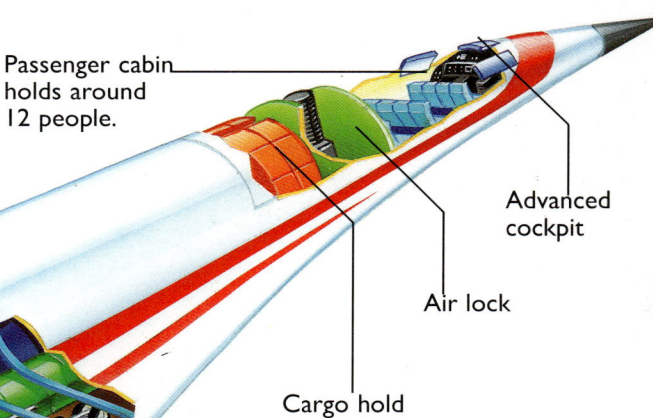

Passenger cabin holds around 12 people.

Advanced cockpit

Air lock

Cargo hold

The plane would use its slower jet engines for landing.

Enormous hydrogen fuel tanks fill most of the plane.

Cruising through space, the high-powered engines could push the plane to speeds as high as 10,000km/h (6200mph).

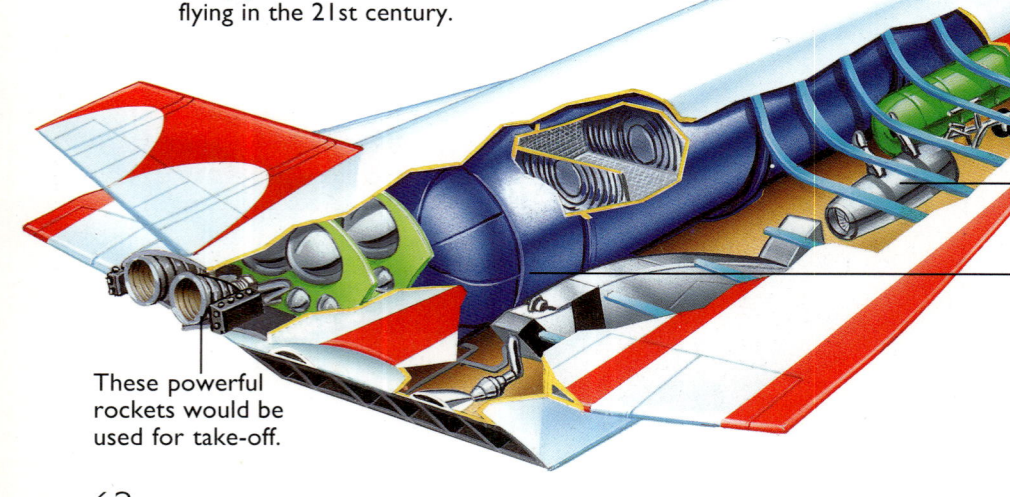

These powerful rockets would be used for take-off.

Pilotless planes

Electronics and computers will become more advanced and control more and more parts of an aircraft. In the future, pilotless planes could shuttle cargo and passengers from one place to another, guided by navigation systems from the ground.

An artist's idea of what a pilotless plane might look like.

Space cargo carriers

Many companies are working on designs for reusable space vehicles to help or replace the Space Shuttle. The design below, from an American company, Lockheed, has no wings. Instead, it relies on its body to create lift. It would take off like a rocket but land like a normal plane on a runway.

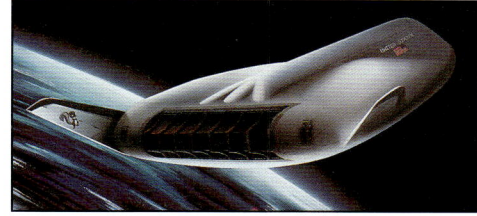

Just over 33m (110ft) long, this craft would carry as much as 500 metric tonnes (550 US tons) of fuel.

Advanced take-off

Many aircraft makers are working on military aircraft with advanced take-off and landing systems. Some are even planning designs for small airliners which could operate from very short runways.

Lockheed's design for a lightweight fighter jet includes a lift fan for very short take-offs and vertical landings.

Glossary

Aerodynamics. The science of how gases, such as air, move over an object. Aerodynamics greatly affect the way planes are designed, built and flown.

Aerofoil. An object shaped to produce lift when air flows over or under it. This shape is most often seen in aircraft wings.

Air pressure. The force with which air pushes against an object. Air pressure is increased by pushing air into a small space. This is called compressing.

Alloys. A metal which is mixed with other metals or substances. For example, steel is an alloy made by mixing iron with carbon. Alloys, particularly of aluminium, are very important in the aircraft industry.

Biplane. An aircraft which has two sets of wings.

Combustion chamber. The part of a gas turbine engine where the air and fuel mixture is set alight.

Compressor. The part of a gas turbine engine where air is squashed together just before it is mixed with fuel and burnt.

Control surfaces. The hinged parts of a plane's wing and tail that help the plane change direction.

Duralumin. A metal alloy (see above) made by adding a small amount of copper to aluminium.

Four forces of flight. The four forces acting on a plane when it flies. The forces are drag, thrust, lift and gravity (see below).

Friction. The resistance made when one surface moves and rubs against another surface and when air moves over a plane.

Fuel management system. A system which adjusts the amount of fuel in each tank and how the tanks supply fuel to the engine or engines.

Galley. The mini kitchen on an airliner where food and drink are prepared for the passengers.

Gravity. A force which pulls objects towards the ground. Gravity has to be overcome by lift from the plane's wings for a plane to rise into the air.

HOTAS. Stands for hands on throttle and stick. Some modern military jets are fitted with an advanced throttle and control column or stick which contains all the controls needed for air combat.

Hydraulics. A system which uses liquid to transmit power from one place to another.

Ignite. To set fire to something.

Inertial Navigation System (INS). An advanced navigation system which measures any changes in the aircraft's speed and direction of movement. The changes are fed into a computer which constantly plots a plane's position. INS is very accurate over long distances.

Mach 1. Mach is a measure of speed. Mach 1 is the speed at which sound travels.

Magnetic field. The area around a magnet that responds to the magnet's power to attract or repel.

Monoplane. An aircraft which has one set of wings.

Pressurization. Keeping the inside of a plane filled with air at a greater air pressure (see above) than the air outside of the plane. This is so that the aircrew and passengers can breathe normally.

Range. The distance an aircraft can travel without running out of fuel.

Stalling. The break up of airflow over the wings caused by not flying fast enough or angling the nose upwards too steeply. Unless corrected, stalling will cause the aircraft to dive dangerously.

Streamlining. To shape an object in a way that makes it move as smoothly as possible through the air. The more easily an aircraft can move, the less power it has to use.

Triplane. An aircraft with three sets of wings.

Turbines. The spinning blades in a gas turbine engine.

Planes: Facts

1783 Invention of the **hot-air balloon** by Joseph and Étienne Montgolfier, two French brothers. The **first hydrogen balloon** was released two months later by another Frenchman, J. Charles.

1804 The English scientist Sir George Cayley built a model glider which is regarded as the **first real plane**.

1852 Henri Giffard from France made the **first controlled, powered flight in an airship** over Paris.

1853 Sir George Cayley built a **glider**, the first heavier-than-air machine to carry a person.

1890 Clément Ader, a French designer, was the **first man to fly in a machine powered by its own engine**.

1890s Otto Lilienthal, a German known as "The Flying Man", made **more than two thousand glider flights**.

1900 Ferdinand von Zeppelin from Germany built the **first successful dirigible airship**. A dirigible is an airship which can be propelled and steered.

1903 Orville and Wilbur Wright made the **first flight in a powered and controlled aircraft** called the *Flyer*. Aviation as we know it began with the Wright Brothers' first powered flights near Kitty Hawk in North Carolina, USA (see page 34).

1905 The Wright Brothers first flew their *Flyer III*. They managed to travel 39km (24 miles). This was the **first efficient plane**. It was able to turn and circle with ease and could fly for over half an hour.

1909 The **first flight across the English Channel** was achieved by a Frenchman, Louis Blériot. He designed and built the plane himself. Blériot was one of the **first to build a successful monoplane** (an aircraft with only one pair of wings).

1910 Henri Fabre, a Frenchman, built a **sea plane**, the *Hydravion*, the **first aircraft to take off from water**.

1913 Igor Sikorsky, a Russian engineer, **built and flew the first four-engined aircraft**.

1913 New, faster airplanes were designed with the advent of air-races during this decade. Adolphe Pégoud was the **first man to loop the loop** and invented what we now call aerobatics (see page 49).

1919 The **first trans-Atlantic crossing**. Lieutenant Commander Read of the US Navy piloted a flying boat, *N-C-4*, from New York, USA to Plymouth, U.K.

1919 John Alcock and Arthur Brown made the **first non-stop trans-Atlantic flight** in a twin-engined Vickers Vimy biplane (a plane that has two sets of wings, one above the other). They flew 1,936 miles from Newfoundland to Ireland.

1919 G. Scott commanded the **first two-way trans-Atlantic flight** from Scotland to New York in the British airship R-34.

1919 **Aircraft and Travel Ltd.** started the **first regular international airline service** between London and Paris, using converted biplane bombers.

1923 The *Autogyro* by Juan de la Cierva made its first flight in Madrid. This was the **first practical aircraft with a rotary wing**, foreshadowing the helicopter.

1926 The **first successful flight over the North Pole** was in a Fokker monoplane by Richard Byrd of the US Navy and Floyd Bennet.

1927 The **first solo, non-stop trans-Atlantic flight** was by Charles Lindbergh in his monoplane, *Spirit of St. Louis*. The journey from Roosevelt field, near New York, to an airport near Paris took 33 hrs 30 mins.

1928 The Australian pilot Sir Charles Kingsford Smith and his crew made the **first trans-Pacific flight** in a Fokker monoplane called the *Southern Cross*.

1930 Frank Whittle, a British engineer, patented a **design for a jet aircraft engine**. Amy Johnson became the **first woman in the world to fly solo** from England to Australia in a second-hand Gypsy Moth.

1932 Amelia Earhart made the **first solo trans-Atlantic flight by a woman** in a Lockheed Vega monoplane.

1933 Wiley Post, an American, was the **first person to fly around the world**.

1933 The **Boeing 247** was built. It was a *streamlined**, all-metal construction. This airplane was the first of the new generation of US domestic air transports.

1936 The Focke-Achgelis (called the FA 61) flew. This was the **first practical helicopter**.

1939 The **first jet-propelled airplane**, the Heinkel *He 178*, was designed by a German, Ernst Heinkel.

1939 The **first successful flight in a single-rotor helicopter** by Igor Sikorsky. He also built several flying boats.

1944 The world's **first production jet aircraft**, *Me 262*, entered service with a maximum speed of just over 800 kph/ 500 mph.

1944 The Germans began to attack London with *V-1* **flying bombs**. The *V-1s* were powered by a type of jet engine known as the pulse jet and could fly without a pilot.

1947 The **first supersonic** (faster than sound) **flight** was successfully completed by an American, Charles Yeager, in a rocket-powered aircraft called the Bell X-1.

1947 The **largest plane ever made**, the *Spruce Goose*, made its only flight and travelled one mile (1.6km). It had eight engines and a wingspan the length of a soccer field.

1949 The **first jet airliner**, the De Havilland Comet, flew.

1953 The **first flight in a VTOL (Vertical Take Off and Landing) plane** (see page 52).

1958 The **first trans-Atlantic jet airline service** was introduced.

1965 The **first flight of a spy plane**, the *SR-71 Blackbird*, built in the United States. It could fly very high and at great speed but has not been used since 1990.

1968 The **first flight of the Russian Tu 144 supersonic airliner**. It was designed to carry 120 passengers at about 1,550 mph.

1969 The **first flight of the Boeing 747 "Jumbo Jet"**. This aircraft is 59m (195ft) long and has over 400 seats (see pages 47 and 50). The **first flight of the airliner Concorde 001**.

1970 The **Boeing 747 "Jumbo Jet"** began **regular passenger service**.

1976 **Concorde entered full-time service** and is the only supersonic plane (see pages 46-47).

1989 The **first flight of stealth plane Northrop B-2 Spirit**. It can fly 8,000km (5,000 miles) in one trip (see page 60).

1991 The **Airbus A340 made its first flight** (see pages 39 and 40). This was built by a group of European companies and can travel from London to Chicago and back before it needs more fuel.

THE USBORNE BOOK OF CUTAWAY BOATS

Christopher Maynard

Designed by
Isaac Quaye & Steve Page

Illustrated by: Mick Gillah, Sean Wilkinson, Ian Cleaver, Gary Bines, Justine Peek and artists from the School of Illustration, Bournemouth and Poole College of Art and Design.

Consultants: Guy Robbins & David Topliss, National Maritime Museum, London

Edited by Jane Chisholm

Additional designs by Robert Walster

Usborne Publishing wish to thank the following for their help with this book:

Joan Barrett, Atlantic Container Line · Alan L. Bates · Historic Royal Dockyard, Chatham · Michael Leek, Cordwainer's College · Flarecraft Inc. · Howard Smith (London) Ltd · Incat Designs Ltd., Japan Ship Centre · National Motorboat Museum · O.Y. Nautor Ab. · P & O · P & O European Ferries · Princess Cruises · Royal National Lifeboat Institution · Eamon Holland, Strategic Advertising · Vasa Museum, Stockholm · Yamaha Motor (UK) Ltd. · Yanmar Diesel Engine Co., Ltd.

Every effort has been made to trace and acknowledge ownership of copyright.
The Publisher will be glad to make suitable arrangements with any copyright holder whom it has not been possible to contact.

Contents

Introduction	66
Triremes	68
The Age of Sail	70
Steamships	72
Riverboats	74
Yachts	76
Ferries	78
Engines	80
Lifeboats	82
Racing boats	84
Container ships	86
Tugs	88
Cruise ships	90
Submersibles	92
The future	94
Glossary	95
Index	96

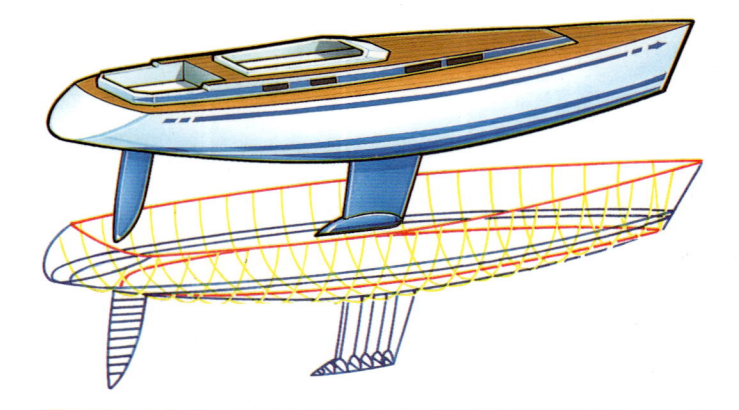

Words in *italic* type

Words in this section which appear in *italic* type and are followed by a small star (for example, *knots**) can be found in the glossary on page 95.

Introduction

Boats have been around since Stone Age times. The earliest boats - dugout canoes, log rafts and frames of sticks covered in animal skins - were all paddled or rowed.

The Ancient Egyptians, in around 3000BC, were the first to use sails to harness the wind. By doing so, they discovered a much easier way to travel, and created enough power to drive much bigger boats.

But really large ships need engines to power them, and it's only in the past 200 years that these have come into use. Some run on diesel, but the fastest and biggest ships use gas or steam turbines.

A tug boat

This is an old tug that was used about 70 years ago to tow ocean-going ships in and out of port. Its power came from a large steam engine that ran on coal.

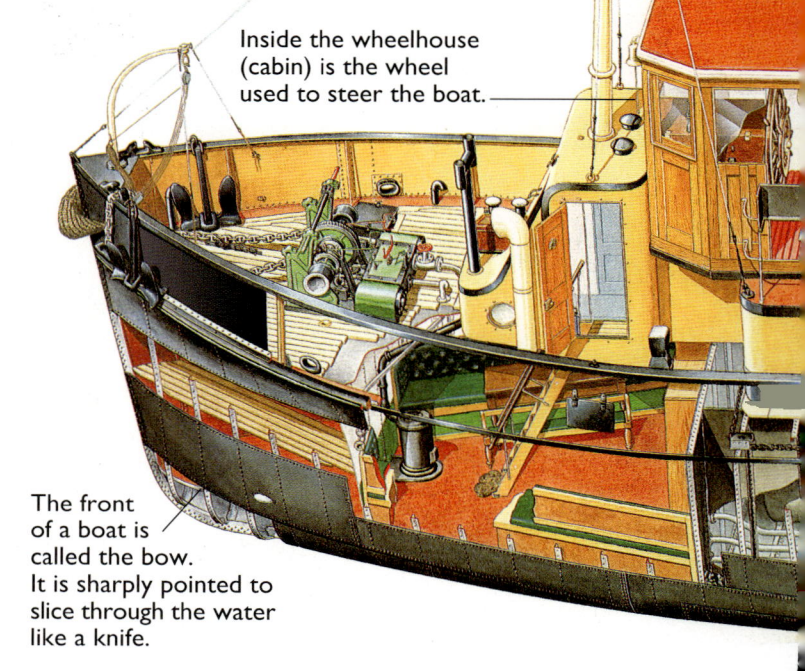

Inside the wheelhouse (cabin) is the wheel used to steer the boat.

The front of a boat is called the bow. It is sharply pointed to slice through the water like a knife.

How boats float.

When a boat is placed in water, it pushes water aside, or displaces it. The water pushes back with a force called upthrust. The size of the upthrust depends on the weight of water displaced. In order to float, an object must displace enough water so that the upthrust is as great as the weight of that object.

The amount of water an object displaces depends on its shape. For example, a ball of clay will sink, but if you hollow it out into a bowl shape it will float. By changing the clay's shape, you have increased the amount of water it displaces. This is what boat builders do. A solid lump of steel would sink, but a ship made of hollowed out steel will float.

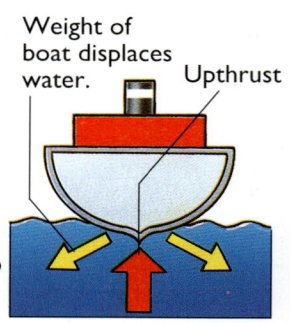

Weight of boat displaces water. Upthrust

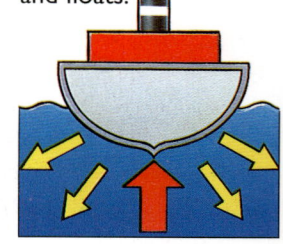

Upthrust equals weight of boat. Boat settles and floats.

How boats steer

A boat is steered by a rudder or a steering oar, which is a big blade-shaped object at the stern (the back of the boat). This cuts into the flow of water and can swivel to deflect the water to either side. As the water pushes hard against the blade, it causes the stern to swing around, pointing the bow of the boat in a new direction.

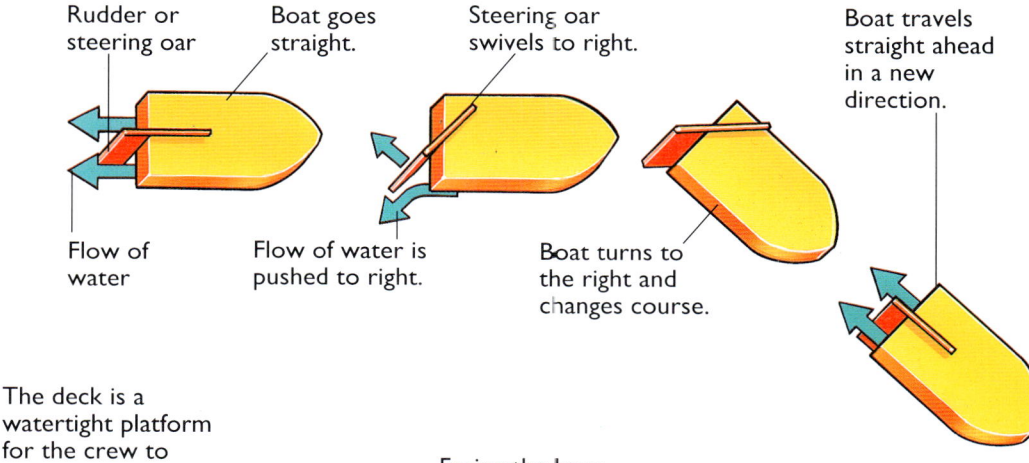

Rudder or steering oar • Boat goes straight. • Steering oar swivels to right. • Boat travels straight ahead in a new direction.

Flow of water • Flow of water is pushed to right. • Boat turns to the right and changes course.

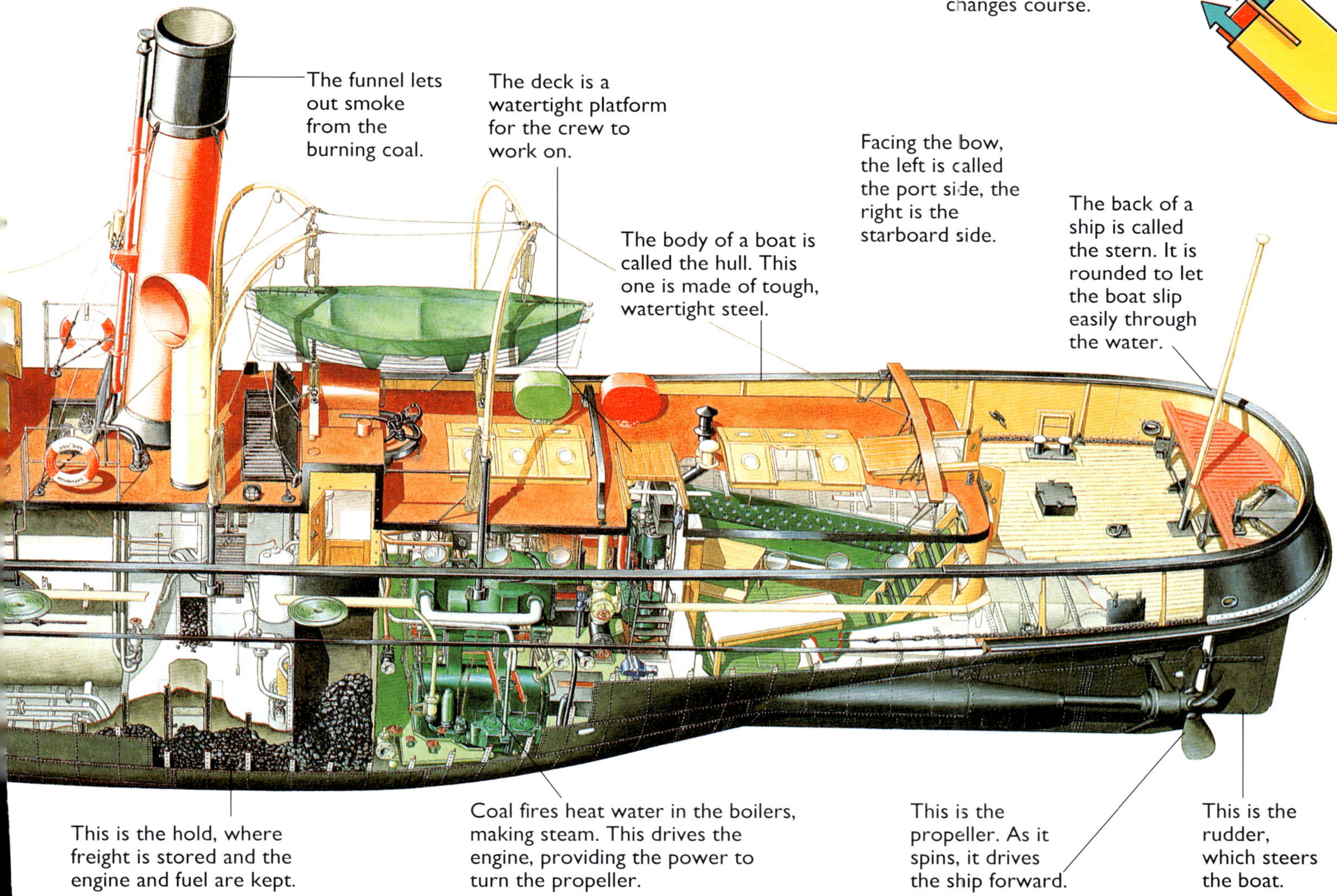

- The funnel lets out smoke from the burning coal.
- The deck is a watertight platform for the crew to work on.
- The body of a boat is called the hull. This one is made of tough, watertight steel.
- Facing the bow, the left is called the port side, the right is the starboard side.
- The back of a ship is called the stern. It is rounded to let the boat slip easily through the water.

- This is the hold, where freight is stored and the engine and fuel are kept.
- Coal fires heat water in the boilers, making steam. This drives the engine, providing the power to turn the propeller.
- This is the propeller. As it spins, it drives the ship forward.
- This is the rudder, which steers the boat.

How boats sail

Sails pointing into the wind • Wind direction • Wind does not push or pull on the sail at all.

Sails sideways to the wind • Wind creates lift at the front of the sail and pulls the boat forward. • Wind pushes on back of sail and shoves boat along.

Sails too far into the wind • No pull on front of sail • Wind still pushes against sail.

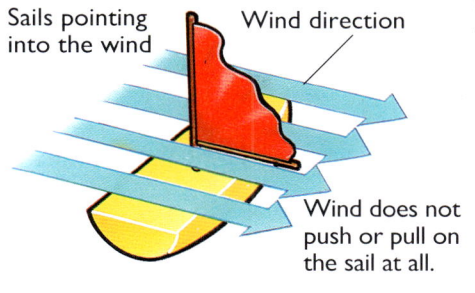

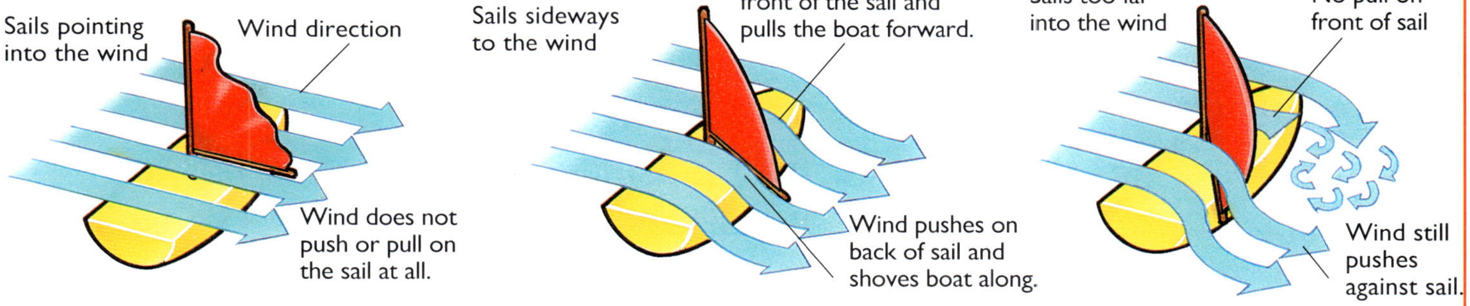

A boat moves by trapping the wind in its sails. But if the sails point directly into the wind, they only flap noisily, producing no power.

Sideways to the wind, the sail fills and creates two forces: lift, which pulls the boat forward, and push, which shoves the boat along.

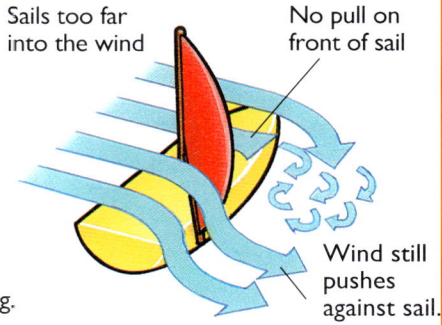

But if a sail is hauled too far into the wind, the airflow behind it breaks up and stops pulling. The sail loses lift and produces much less power.

Triremes

About 2,500 years ago, in Ancient Greece, the most powerful and famous warship in the world was the trireme. It was big, fast and deadly, even though it was always rowed into battle.

As Greek cities grew rich and powerful, fleets of triremes were built to patrol the waters of the eastern Mediterranean. These ships cost a great deal to run, so only the richest cities, such as Athens or Corinth, could afford very many of them.

Olympias

In 1985, a group of ship lovers and historians from all over the world launched a full-size replica trireme. It was called *Olympias* and, fittingly, it was built in Greece.

The main mast and the foremast each had one sail. The sails were only used on longer journeys when the wind was in the right direction.

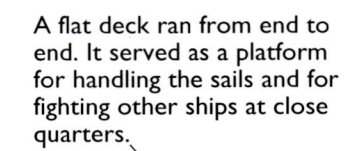

A fast-moving ship might pack a punch of 60 tonnes (58.8 tons) or more as it hit another boat.

A flat deck ran from end to end. It served as a platform for handling the sails and for fighting other ships at close quarters.

Triremes were so long and narrow that cables of rope from bow to stern were needed to stiffen them. Otherwise they would have drooped at either end.

The bow ended in a 2m (6.5ft) wooden ram fitted with a heavy jacket of bronze. Rams were used to punch holes in the hulls of enemy ships.

The bow was decorated with a painted eye to scare the enemy.

A sunken gangway down the middle of the deck let rowers climb in and out of their seats.

The engine room

Three banks of rowers were the engine that drove a trireme. They all used incredibly long oars, and could drive a ship at an amazing speed of over eight *knots** (almost 15 km or 9 miles an hour) all day long. This was much faster than it could travel by sail. When going a long distance in a hurry, triremes were almost always rowed. A long day's voyage from dawn to dusk might cover as much as 220km (136 miles).

Triremes floated at a depth of 1m (just over 3ft). Being so shallow meant they could sail very close to shore and haul up onto a beach.

The upper rowers were called thranites. They sat in two lines of 31.

Middle rowers were called zygites. They sat in two rows of 27.

On the bottom tier were the thalamites. They sat in rows of 27, too.

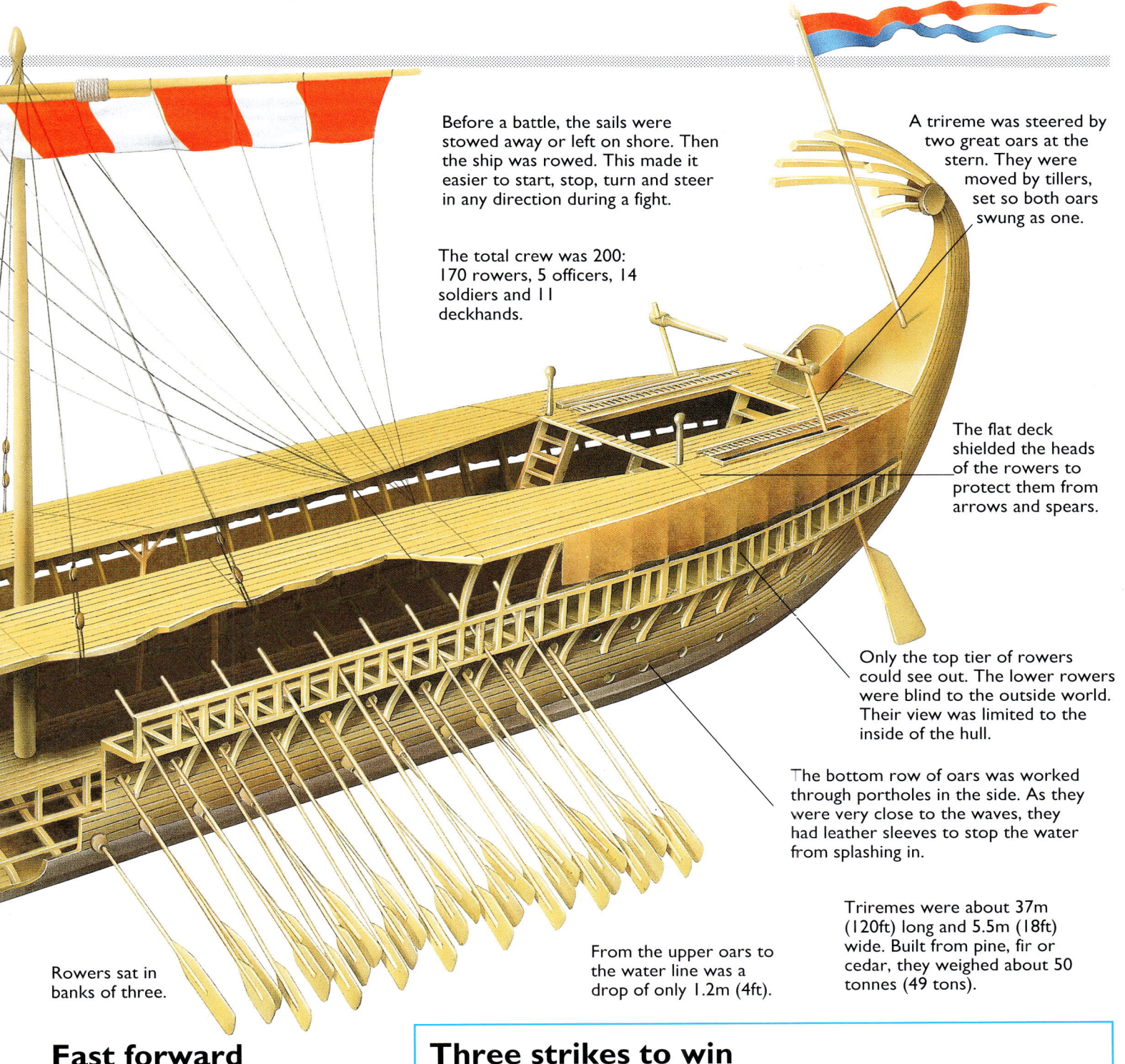

Before a battle, the sails were stowed away or left on shore. Then the ship was rowed. This made it easier to start, stop, turn and steer in any direction during a fight.

The total crew was 200: 170 rowers, 5 officers, 14 soldiers and 11 deckhands.

A trireme was steered by two great oars at the stern. They were moved by tillers, set so both oars swung as one.

The flat deck shielded the heads of the rowers to protect them from arrows and spears.

Only the top tier of rowers could see out. The lower rowers were blind to the outside world. Their view was limited to the inside of the hull.

The bottom row of oars was worked through portholes in the side. As they were very close to the waves, they had leather sleeves to stop the water from splashing in.

From the upper oars to the water line was a drop of only 1.2m (4ft).

Triremes were about 37m (120ft) long and 5.5m (18ft) wide. Built from pine, fir or cedar, they weighed about 50 tonnes (49 tons).

Rowers sat in banks of three.

Fast forward

Trireme oars were long - 4.3m (14ft) from tip to handle. That's far higher than the ceiling of most modern rooms. The Ancient Greeks knew that the blade of a long oar swept much farther - and more powerfully - through the water than a short oar.

A Greek rower making a single stroke of his oar.

Distance covered by a short oar blade

Distance covered by a long oar blade

Three strikes to win

At the start of battle, triremes often faced off in two long lines. Each ship would pick out a target, then dart forward to try to ram an opponent and sink him. The best places to aim for were the stern and sides of another ship.

A trick often used was to sweep around the far end of the opposing line of ships and strike from the rear.

A daring move was to break through a gap in the line, wheel and strike from behind.

Sometimes a ship made for a gap in the line, then veered at the last moment to smash into the side of an enemy with its ram and shear off its oars.

69

The Age of Sail

Rowing is fine for lightweight boats, but it takes a lot more power to drive a really big ship through the water. For thousands of years, people relied on the wind. Using masts and sails, they were able to harness its energy to propel big ships all over the world.

The Vasa

The *Vasa*, launched in 1628, was the pride of the Swedish navy. But, because of a faulty design, she sailed a very short distance before sinking. Raised from the sea in 1961, she is the only complete 17th century warship in the world.

The Vasa
- Length: 69m (226ft)
- Width: 12m (39ft)
- Height: 53m (174ft)
- Weight: 1,300 tonnes (1,274tons)
- Guns: 64 guns
- Crew: 135 sailors
       300 soldiers

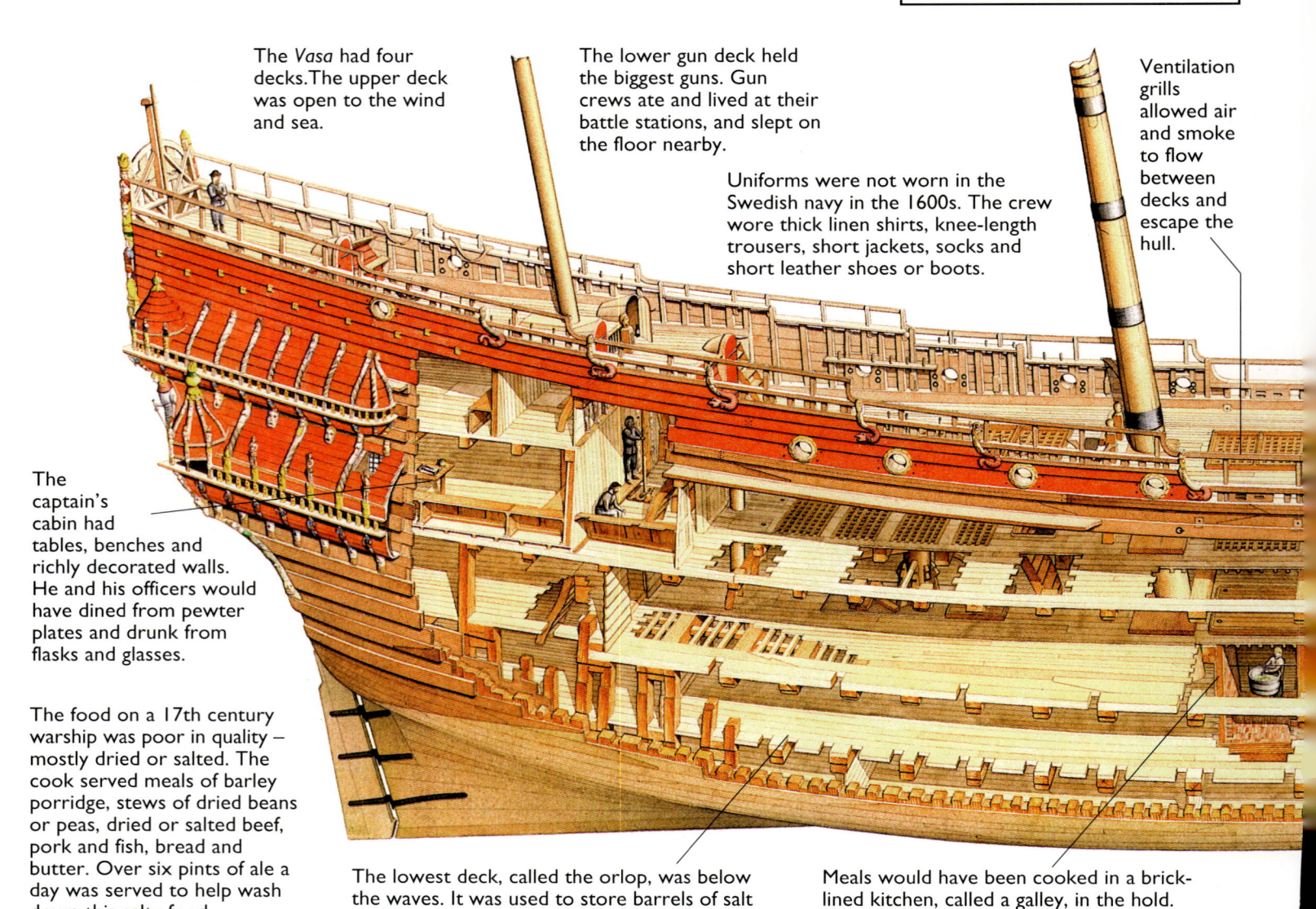

The *Vasa* had four decks. The upper deck was open to the wind and sea.

The lower gun deck held the biggest guns. Gun crews ate and lived at their battle stations, and slept on the floor nearby.

Uniforms were not worn in the Swedish navy in the 1600s. The crew wore thick linen shirts, knee-length trousers, short jackets, socks and short leather shoes or boots.

Ventilation grills allowed air and smoke to flow between decks and escape the hull.

The captain's cabin had tables, benches and richly decorated walls. He and his officers would have dined from pewter plates and drunk from flasks and glasses.

The food on a 17th century warship was poor in quality – mostly dried or salted. The cook served meals of barley porridge, stews of dried beans or peas, dried or salted beef, pork and fish, bread and butter. Over six pints of ale a day was served to help wash down this salty food.

The lowest deck, called the orlop, was below the waves. It was used to store barrels of salt beef and pork and other dried food. Sails, ropes and spare equipment was kept here too.

Meals would have been cooked in a brick-lined kitchen, called a galley, in the hold. A cauldron hung over an open fire. Smoke flowed freely up to the decks above.

Packing a big punch

Vasa was one of the most powerful ships of the Swedish Navy. It carried 64 guns, including 48 big ones able to fire 11kg (24 lb) cannonballs. Together they weighed over 72 tonnes (71 tons).

The *Vasa*'s guns were the most high-tech weapons of their time. But they were slow. Ten rounds an hour was considered outstanding. Between each firing, the gun had to be cleaned and left to cool.

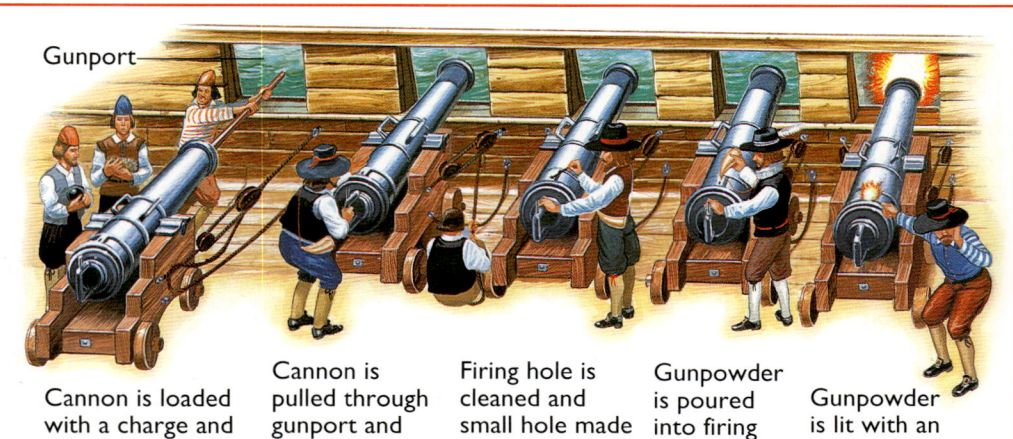

Gunport

Cannon is loaded with a charge and cannonball.

Cannon is pulled through gunport and aimed.

Firing hole is cleaned and small hole made in main charge.

Gunpowder is poured into firing hole.

Gunpowder is lit with an explosive fuse.

Rigged for sailing

The *Vasa* was a three-masted ship. She could put up ten sails in all, and a flutter of pennants and flags. At the time of sinking she was flying four sails. The other six were still in lockers. Today they are preserved intact in the Vasa Museum in Stockholm.

Royal warship

The *Vasa* was a fighting ship, but she was also built to show off the wealth and power of the King of Sweden, Gustavus Adolphus. From top to bottom, the entire stern was richly carved with hundreds of gilded figures and ornaments, including a huge royal coat of arms flanked by two crowned lions. Even the hatch covers of the gunports had faces of roaring lions carved onto them.

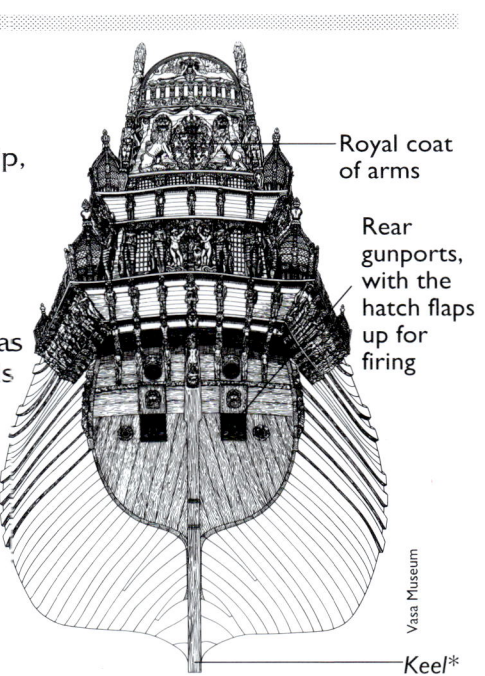

- Royal coat of arms
- Rear gunports, with the hatch flaps up for firing
- Keel*

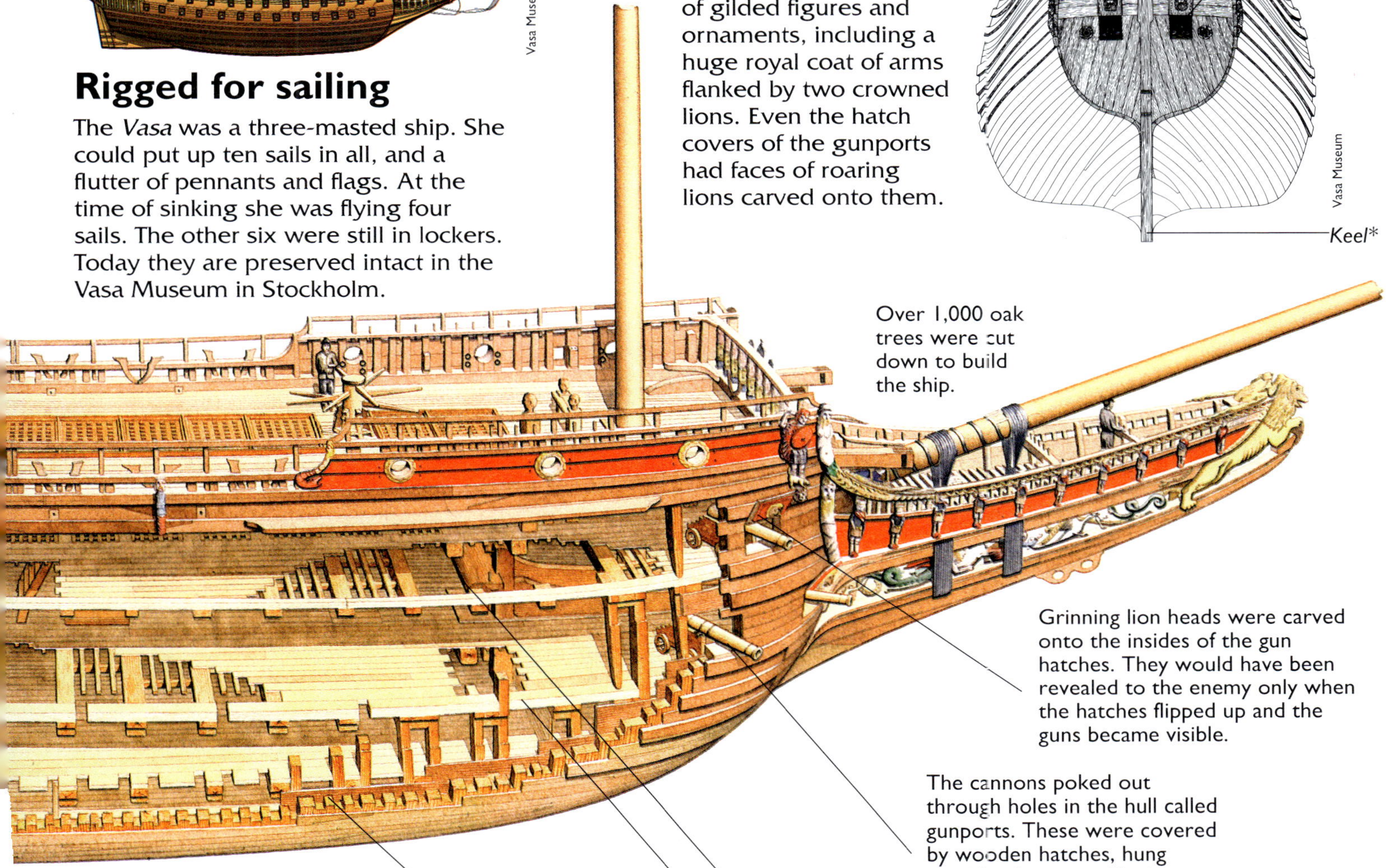

Over 1,000 oak trees were cut down to build the ship.

120 tonnes (118 tons) of stone *ballast** were packed in the hold to balance the weight of the masts and sails.

Gunpowder was stored in the hold, well below the water line.

The gun decks were dark, damp and crowded. The ship had no heating to keep out the chill.

Grinning lion heads were carved onto the insides of the gun hatches. They would have been revealed to the enemy only when the hatches flipped up and the guns became visible.

The cannons poked out through holes in the hull called gunports. These were covered by wooden hatches, hung outside, that were lifted by ropes when the guns were ready to fire.

Why did the *Vasa* sink?

The *Vasa* sank because she was top-heavy. She was built too big and strong, and had too many heavy guns on the deck for the size of *hull**. She was also far too narrow to carry all that weight above the water line and still keep her balance. Just a mild gust of wind was enough to overturn her.

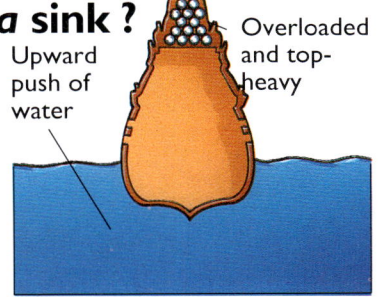

- Upward push of water
- Overloaded and top-heavy

Although the *Vasa* was heavy, she could still float, because the upward push of water was equal to her weight. But being top-heavy made her unstable.

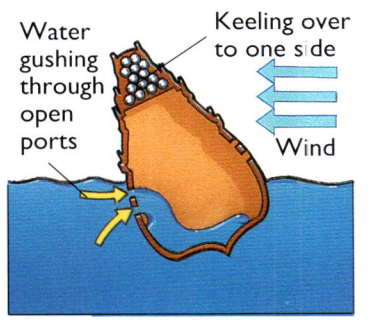

- Water gushing through open ports
- Keeling over to one side
- Wind

The ship began to roll heavily in the breeze and a sudden gust of wind made her lean sharply to one side. Then water began to flood in through open gunports.

Ship fills with water and sinks.

The water flooding in added extra weight to the ship, overcoming the upward push of the water below. The ship sank like a stone.

Steamships

In the 1800s, steam engines began to be installed in ships powered by sail, like *H.M.S. Gannet* shown here. This new source of power enabled a ship to travel without being dependent on winds or tides. With an engine to drive a propeller, it could make headway even in complete calm.

H.M.S. Gannet was a three-masted ship built by the British Navy to protect the sea routes of the empire.

Switching to sail

When the *Gannet* wanted to sail, the engine was shut down. The funnel - which lowered like a telescope - was dropped and the sails were hoisted.

As the propeller (also known as the screw) now slowed the ship down, it was unhooked and lifted out of the water. The crew used a big deck winch and chain to raise it.

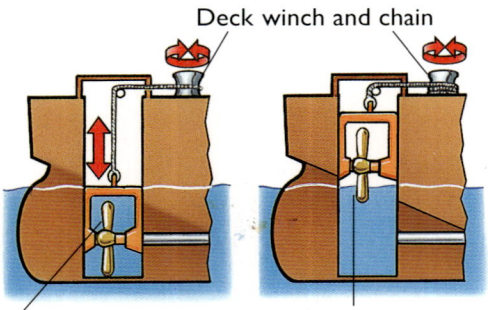

Deck winch and chain

Propeller unhooked   Propeller lifted up

The ship had a crew of 140 men and boys.

Foredeck

H.M.S Gannet was fitted with six medium guns that fired shells of solid steel, and six to eight machine guns.

At full speed, the engine could drive the ship at 11.5 knots. Yet under sail the ship went even faster - sometimes as much as 15 knots.

Every corner of the hold was stuffed with equipment intended to last for two or three years. Spare parts for the ship were almost impossible to find in the regions to which she sailed.

The *Gannet* carried over 142 tonnes (140 tons) of coal in her bunkers, enough to travel more than 1,600km (994 miles).

Ten iron bulkheads (walls with watertight doors) divided up the hull. They prevented the whole hull from flooding if any part of it was holed.

The hull was built as an iron frame. A double layer of thick teak planks was bolted onto it.

Steam engines

Steam ships are powered by engines which have boilers and furnaces to produce steam. Once the steam is at high pressure, it is piped to a small cylinder. It then flows on to a large cylinder at lower pressure. These two cylinders drive the pistons that turn the propeller that drives the ship forward.

Boiler containing water

Coal-burning furnace

Water boils and turns into steam.

Steam rushes into small high-pressure cylinder.

Small high-pressure cylinder

Steam forces piston back and forth.

Steam cools and loses some pressure.

Steam enters the larger cylinder and sets the second piston moving.

Propeller

The pistons turn the shaft that drives the propeller.

Large low-pressure cylinder

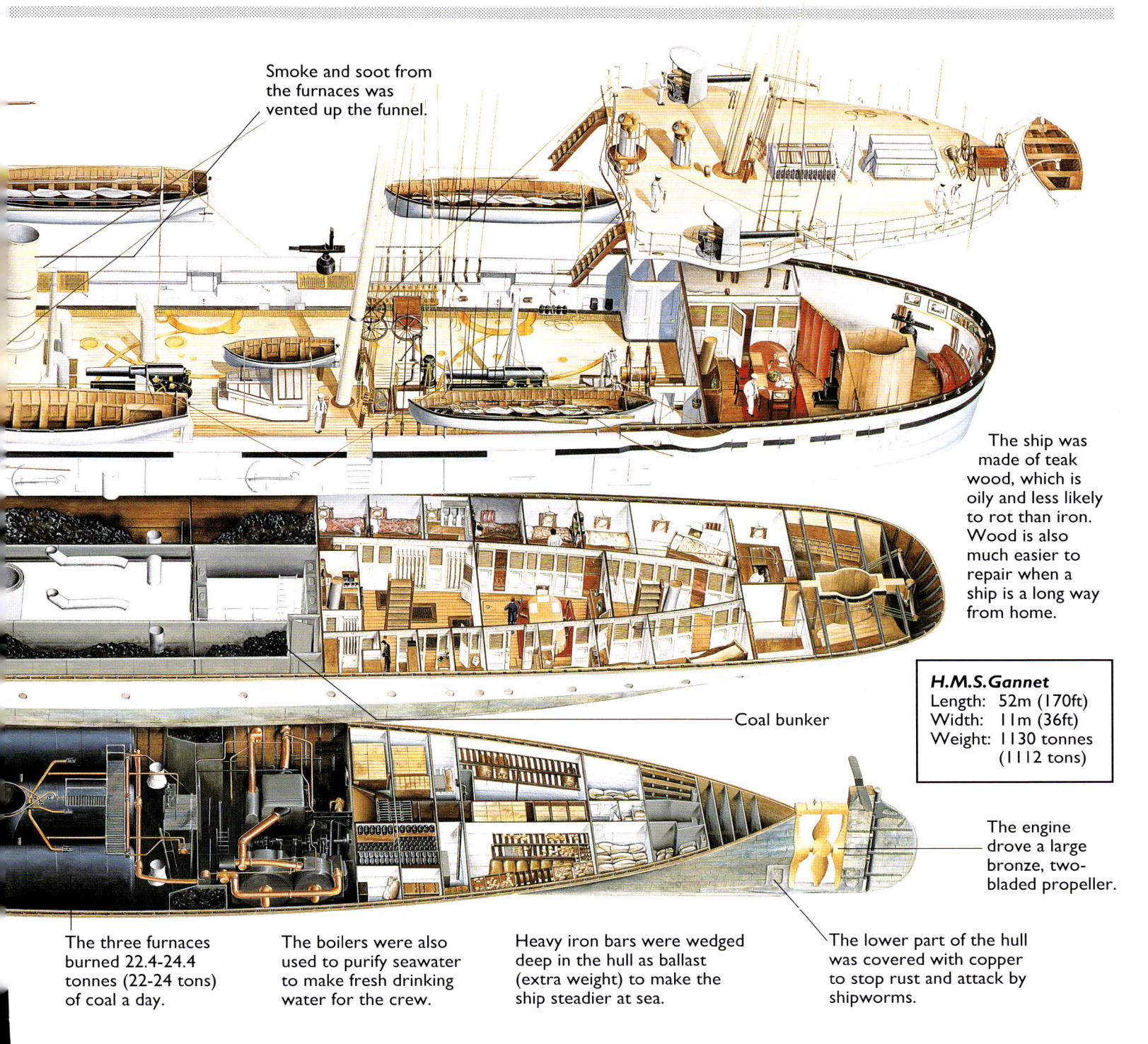

Smoke and soot from the furnaces was vented up the funnel.

The ship was made of teak wood, which is oily and less likely to rot than iron. Wood is also much easier to repair when a ship is a long way from home.

Coal bunker

H.M.S. Gannet
Length: 52m (170ft)
Width: 11m (36ft)
Weight: 1130 tonnes
(1112 tons)

The engine drove a large bronze, two-bladed propeller.

The three furnaces burned 22.4-24.4 tonnes (22-24 tons) of coal a day.

The boilers were also used to purify seawater to make fresh drinking water for the crew.

Heavy iron bars were wedged deep in the hull as ballast (extra weight) to make the ship steadier at sea.

The lower part of the hull was covered with copper to stop rust and attack by shipworms.

H.M.S. Gannet's engine and propeller

Weight:	45 tonnes (44.3 tons)	Steam:	27kg (60lbs)
Length:	4.8m (15ft)	Power:	1100 *horsepower**
Cylinders:	2	Propeller width:	4m (13ft)
Boilers:	3	Propeller weight:	16 tonnes (15.7 tons)
		Propeller speed:	100 revolutions (turns) per minute

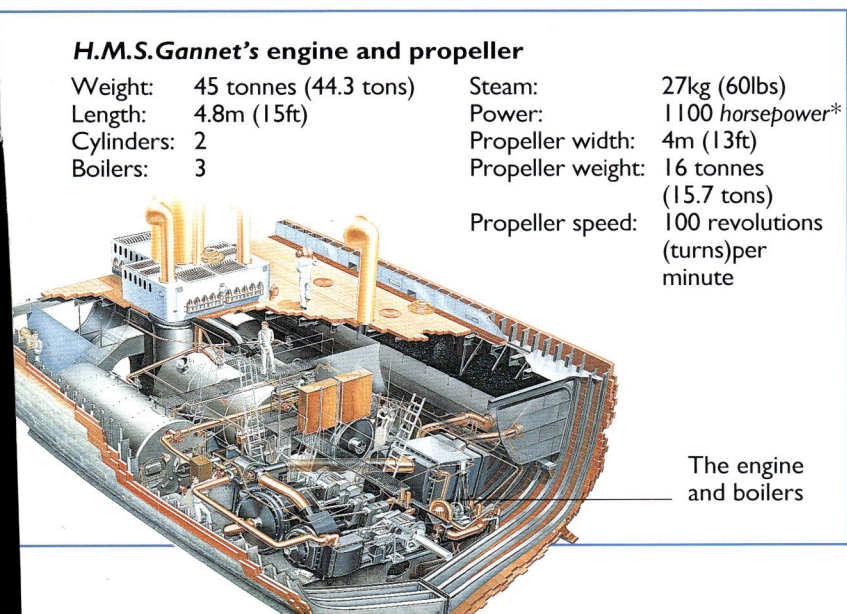

The engine and boilers

How fast is a knot?

A ship's speed is measured in knots. One knot is the same as 1.8km/h (just over a mile an hour). The word comes from the old custom of throwing a knotted rope, tied to a small plank, over the bow. As the wood floated toward the stern, knots in the rope were counted out to calculate the ship's speed.

Riverboats

The riverboat fleet that plied the Mississippi River basin in the 1800s was everything that trucks, trains and planes are today. For years it was the main form of transportation in the region. In 1860, a total of 10 million tons of cargo was shipped this way.

Riverboats that carried passengers and freight were called packets. The biggest were stately palaces that ran long distance express services. They were lightly built, flimsy even, compared to ocean-going ships. They had flat, shallow hulls, since anything deeper than 1.5m (5ft) really limited the places they could get to. But they all had huge steam engines to battle upstream against fast-flowing currents.

The Rob't. E. Lee

The *Rob't. E. Lee*, named after the Confederate commander-in-chief of the southern troops in the American Civil War, was the most celebrated riverboat of all. Built in Indiana in 1866, for the next ten years she worked the Mississippi up and down from New Orleans.

The Rob't. E. Lee
- Length: 87m (285ft)
- Width: 14m (46ft)
- Engines: 2 steam engines
- Boilers: 8 boilers producing 55kg (120lbs) of steam each
- Weight: 1,432 tonnes (1,456 tons)
- Fuel: Coal and wood burning

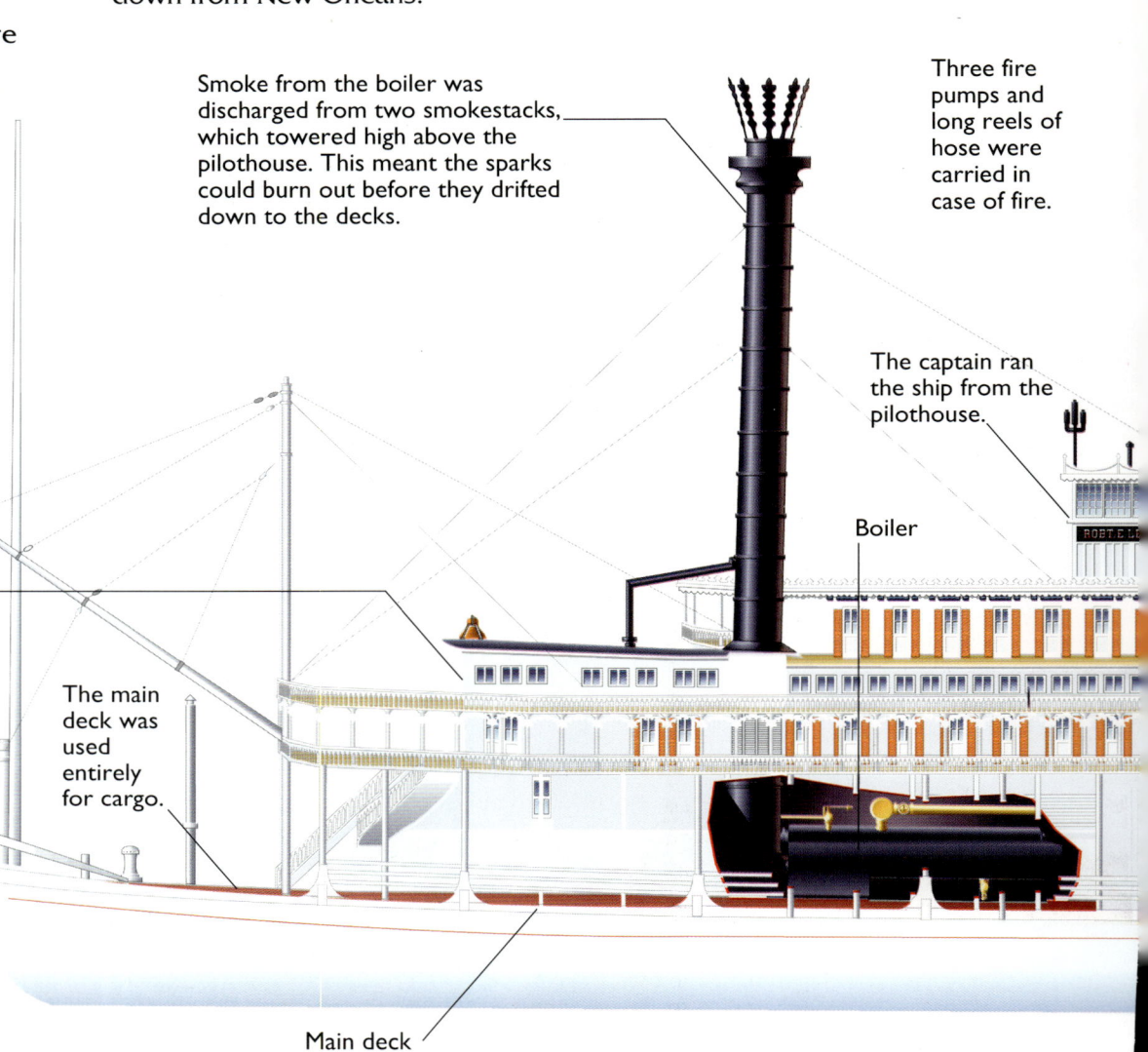

The main cabin was furnished with everything from a velvet carpet to rosewood chairs and sofas.

The entire ship was built of wood, nails, bolts and iron fastenings.

Two steam engines generated 2,700hp* of power, enough to drive her along at over 32 kmph (20mph) in calm waters.

Smoke from the boiler was discharged from two smokestacks, which towered high above the pilothouse. This meant the sparks could burn out before they drifted down to the decks.

The main deck was used entirely for cargo.

Three fire pumps and long reels of hose were carried in case of fire.

The captain ran the ship from the pilothouse.

Boiler

Main deck

Deck cargo

Riverboats took passengers and baggage, but their main business was freight, especially cotton. They carried it from all over the southern states of the USA down to New Orleans to be shipped overseas.

The *Rob't. E. Lee* once loaded 5,741 bales of cotton along the sides of its main deck. They could be stacked one on top of another, past all three decks. Passengers might make a whole trip without ever glimpsing any scenery.

Paddlewheels

Riverboats were powered by huge paddlewheels, mounted on each side, or at the stern. The big advantage over propellers was that they didn't dip below the hull, so the boats could keep going in very shallow waters.

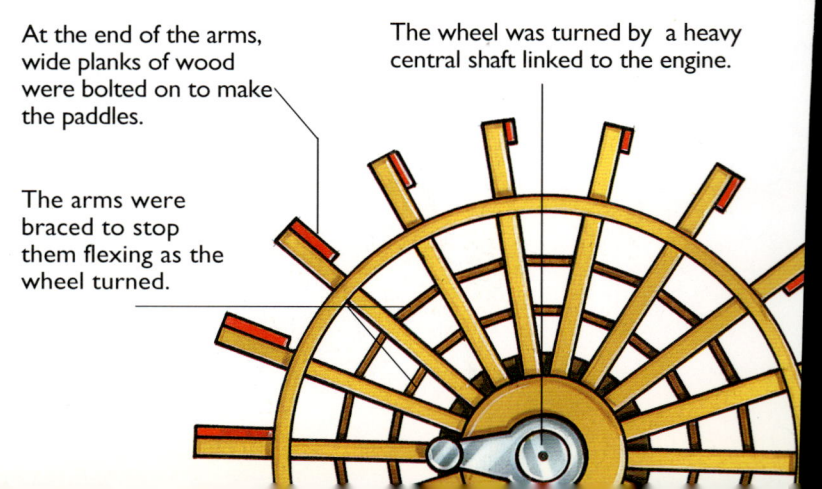

At the end of the arms, wide planks of wood were bolted on to make the paddles.

The arms were braced to stop them flexing as the wheel turned.

The wheel was turned by a heavy central shaft linked to the engine.

Side-wheelers and stern-wheelers

All riverboats had flat, shallow hulls, no keel and did not float very deep in the water.

Side-wheelers had one paddlewheel on each side, and an overhanging main deck which stuck out far beyond the hull. Stern-wheelers had only one wheel at the back, and their main deck was a lot narrower.

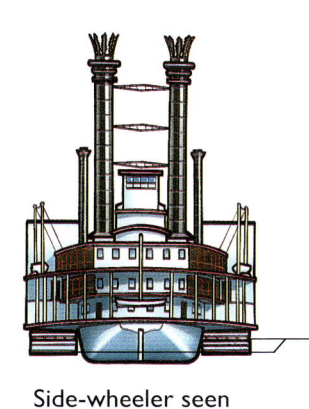

Paddlewheels did not dip below the bottom of the hull. This protected them from rocks, logs and other clutter lying on the bed of the river.

Side-wheeler seen from the bow (front)

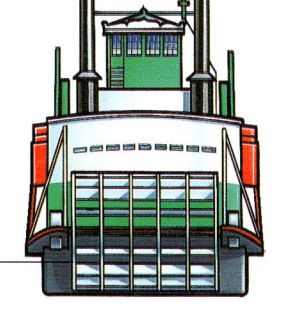

Stern-wheeler seen from the stern (back)

The two main engines had cylinders wide enough for two men to crawl inside.

The middle and top decks were for passenger cabins.

The top deck, known as the Texas deck, had 24 cabins for passengers.

The main deck had 61 staterooms.

Steam engines — Paddlewheel — Rudder

Braces and chains

The hulls of riverboats were so long and thin that they became rather floppy. The bow and stern tended to sag into the water, a habit known as hogging. To correct this, sets of posts and chains were rigged up on deck to stiffen the frame of the hull.

From bow to stern, sets of hog chains were rigged above the deck to stiffen the hull. They were made of lengths of iron rod, screwed together and braced by wooden posts.

Cross-section of hull showing chains from side to side

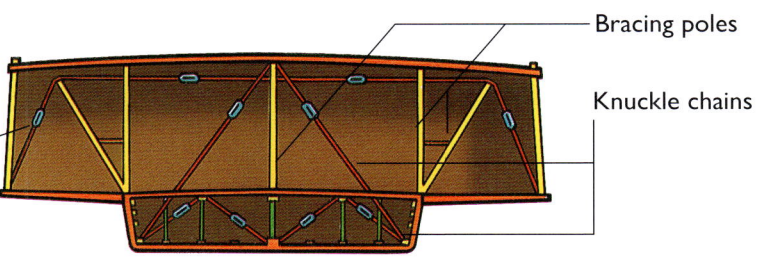

Cross chains — Bracing poles — Knuckle chains

The Great Race

In 1870, the *Rob't. E. Lee* gained lasting fame for itself in a great river race against the *Natchez*.

Both boats were due to leave New Orleans at 5pm on June 30th, bound for St. Louis, and the event grew into a feverishly-awaited race. Hundreds of thousands of dollars were laid as bets, and a huge, excited crowd lined the riverbanks to see the start.

The *Rob't. E. Lee* pulled away a couple of minutes before 5pm, her rival four minutes later. Throughout the lengthy race she was never really challenged again, although a leak in one of her boilers almost put out the fires below before it was finally plugged. The loss of speed let the *Natchez* get within 400 yards for a short while. On the last stage, by chance, night fog let the *Rob't. E. Lee* gain several hours lead. She finally arrived in St. Louis on July 4th, a record 3 days, 18 hours and 14 minutes after setting out (and more than 6.5 hours ahead of the *Natchez*). To this day, no steamboat has ever beaten her time.

St. Louis (Finish line)
Cairo
Memphis
Vicksburg
Mississippi River
Natchez
New Orleans (Start of race)

Yachts

Using sail power today may seem old-fashioned, but modern yachts are very different from their forerunners. Their hulls are made from synthetic materials and super-strong glue, which is far tougher and longer-lasting than wood. The masts are shaped from lightweight metal, which is lighter than wood and doesn't rot in salt water. Many yachts are equipped with quiet diesel engines, as well as the latest satellite navigation gear, two-way radios and depth finders.

The *Swan 55*

The *Swan 55* is a single-masted yacht, known as a sloop. This yacht is built beside the Baltic Sea, in northern Finland, and is designed to be sailed across oceans. It has a deep, rounded hull and high sides. These features are common to all cruising yachts, making them stable and dry at sea.

Modern sails are made of terylene, a strong material which holds its shape well.

Navigation area

The helmsman steers the yacht from the cockpit.

Small sundeck

The rear deck locker stores gas bottles for the stove, a life raft and ropes.

The rudder steers the yacht. It is moved by wires linked to a wheel in the cockpit. The rudder's long blade digs deep into the water to keep the boat dead on course.

Fully-equipped galley (kitchen)

Locker space between the inside walls and the hull

The three-blade propeller is driven by a six-cylinder, 116hp* diesel engine. In calm waters it can do 10 knots*.

There is a sound-proofed walk-in engine room.

The *Swan 55*
Length: 16.7m (55ft)
Width: 4.8m (16ft)
Weight: 23 tonnes
             (22.5 tons)
Draught: 2.6m (8ft)
Sail area: 125m² (1345ft²)

Different yacht types

Sloop — Jib, Mainmast, Mainsail

Yawl — Jib, Mainsail, Mizzenmast, Mizzensail

Ketch — Jib, Mainsail, Mizzenmast, Mizzensail

Schooner — Jib, Foremast, Foresail, Mainmast, Mainsail

Sailing yachts get their names from the way their masts and sails are rigged. A boat rigged with one mast is known as a sloop.

Yawls and ketches have two masts and can fly three or more sails. A mainsail, mizzensail and jib are the most common.

Schooners have a foremast ahead of the mainmast. They are bigger and can carry more sails than most other kinds of yachts.

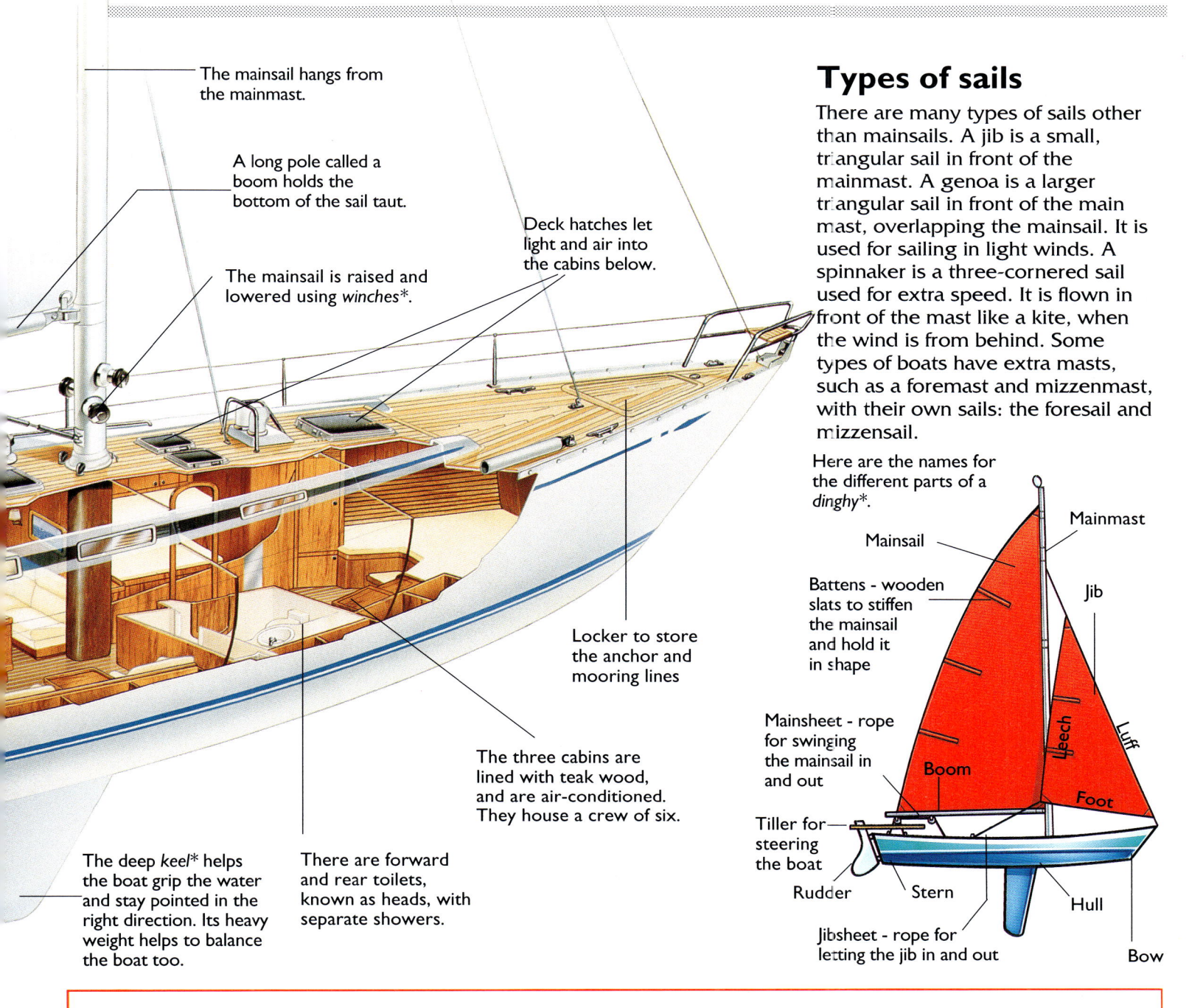

The mainsail hangs from the mainmast.

A long pole called a boom holds the bottom of the sail taut.

The mainsail is raised and lowered using *winches**.

Deck hatches let light and air into the cabins below.

Locker to store the anchor and mooring lines

The three cabins are lined with teak wood, and are air-conditioned. They house a crew of six.

The deep *keel** helps the boat grip the water and stay pointed in the right direction. Its heavy weight helps to balance the boat too.

There are forward and rear toilets, known as heads, with separate showers.

Types of sails

There are many types of sails other than mainsails. A jib is a small, triangular sail in front of the mainmast. A genoa is a larger triangular sail in front of the main mast, overlapping the mainsail. It is used for sailing in light winds. A spinnaker is a three-cornered sail used for extra speed. It is flown in front of the mast like a kite, when the wind is from behind. Some types of boats have extra masts, such as a foremast and mizzenmast, with their own sails: the foresail and mizzensail.

Here are the names for the different parts of a *dinghy**.

Mainmast
Mainsail
Jib
Battens - wooden slats to stiffen the mainsail and hold it in shape
Mainsheet - rope for swinging the mainsail in and out
Boom
Leech
Luff
Foot
Tiller for steering the boat
Rudder
Stern
Hull
Jibsheet - rope for letting the jib in and out
Bow

Sailing and wind direction

Boats can sail in any direction, except straight into the wind, or up to 45° either side of it. Within this area (called the "No Go Zone"), sails flap and lose the power to drive a boat forward. So, to sail into the wind, a boat has to zigzag its way forward. This is known as tacking. Here are three ways of using the wind to sail a boat.

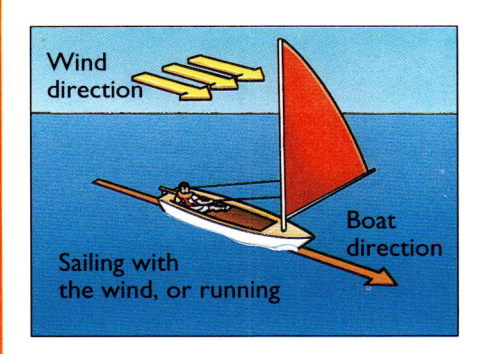

Wind direction
Sailing with the wind, or running
Boat direction

Here, the boat heads in more or less the same direction as the wind, with the sail set at right angles to it. This is a slow way to sail.

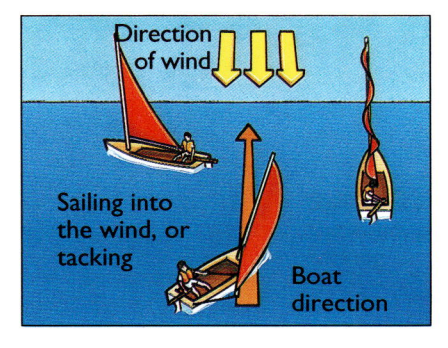

Direction of wind
Sailing into the wind, or tacking
Boat direction

The boat zigzags its way forward at an angle to the wind. It keeps switching the side of the sail that faces the wind to stay on course.

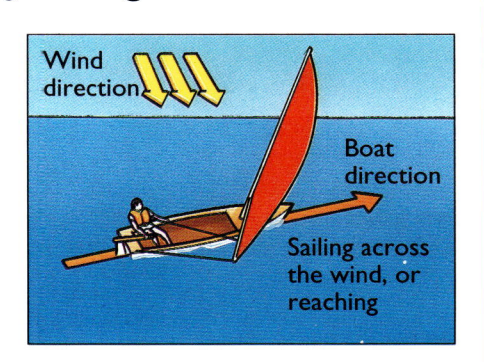

Wind direction
Boat direction
Sailing across the wind, or reaching

Both the sails and the boat lie sideways to the wind. This is a course that traps the wind best and makes for the most speed.

77

Ferries

Some of the busiest ferry routes in the world cross the English Channel. At the height of summer, over 130 trips a day are made between Calais in France and Dover in England, with ships leaving port every 30 minutes.

From inside, the ferries on this short sea journey look like multilevel car parks. Beneath their comfortable passenger areas, they stow hundreds of cars and trucks.

Pride of Calais

On a single trip from Dover to Calais, the *Pride of Calais* superferry can carry up to 2,300 passengers and 650 cars or 100 trucks. The ship works day and night, all year long. It makes the 42km (26 mile) crossing in about 75 minutes, and usually stays in port for less than an hour before its next trip.

One of the busiest lines in the English Channel is P&O European Ferries. It runs five ships from Dover to Calais, including the *Pride of Calais*.

Pride of Calais
Weight: 26,500 tonnes (25,970 tons)
Length: 170m (558ft) (as long as 1½ soccer fields)
Width: 28m (92ft)
Speed: 22 knots
Crew: 110-120 per shift

The *Pride of Calais* and its sister ship the *Pride of Dover* between them carried 5 million passengers in 1995.

Over two million meals a year are served in four different types of restaurants.

Cars and trucks can drive on at one end, park, and then drive straight off at the far end, without ever having to turn or back up.

People in the Club Class lounge have access to phones, faxes, writing desks and photocopiers, so they can work as they travel.

The main diesel engine sits below the car deck.

Electronic eyes

In crowded waters, fishing boats and yachts sail by every day. The heavy traffic keeps a captain alert even in fine weather. But at night, or in fog or storms, the only way to see what's out there is by radar. The long whirling bars at the top of ferry masts are radar antennae. They can see ships, islands, marker buoys and even landmarks on the coast.

1. Radar signals are broadcast from the ferry, some 500 times a second. They travel at the speed of light.

2. If they hit another ship 30 km (18.6 miles) away, a faint echo bounces back.

3. The signals are reflected back to the ferry 1/500th second later.

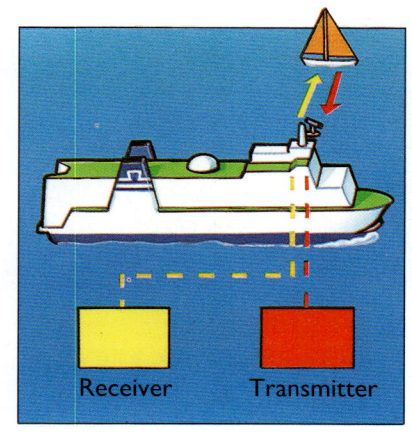

Receiver   Transmitter

Ship on screen

The receiver turns the echo into a bright light on the radar screen. The navigator uses this to track the direction, speed and distance of the ship and steers a course to avoid it.

Roll-on, roll-off

The vehicle decks have wide bow and stern doors, and huge open parking spaces. This means that hundreds of vehicles can drive in and out swiftly with ease, so loading never takes long. This arrangement is known as roll-on, roll-off, or ro-ro.

A two-lane ramp beneath the footbridge loads cars and vans. Lower still is a third ramp for trucks. This ship is loading at the stern. It will unload by the bow.

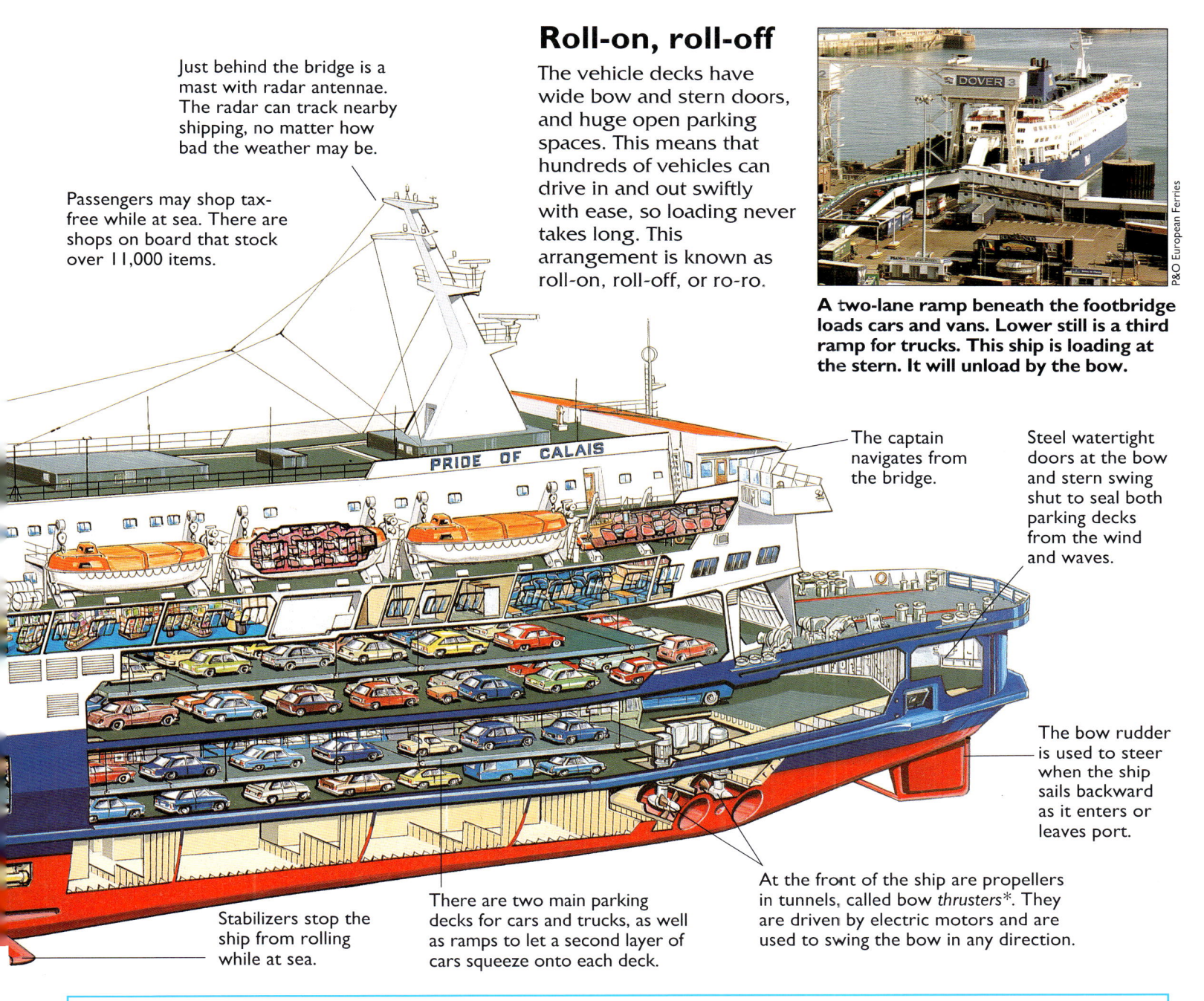

Just behind the bridge is a mast with radar antennae. The radar can track nearby shipping, no matter how bad the weather may be.

Passengers may shop tax-free while at sea. There are shops on board that stock over 11,000 items.

The captain navigates from the bridge.

Steel watertight doors at the bow and stern swing shut to seal both parking decks from the wind and waves.

The bow rudder is used to steer when the ship sails backward as it enters or leaves port.

At the front of the ship are propellers in tunnels, called bow *thrusters**. They are driven by electric motors and are used to swing the bow in any direction.

There are two main parking decks for cars and trucks, as well as ramps to let a second layer of cars squeeze onto each deck.

Stabilizers stop the ship from rolling while at sea.

Air-cushioned vehicles

Ferries that use a powerful cushion of air to lift themselves off the ground are called air-cushioned vehicles, or ACVs. ACVs can hover in one place, or move forward, backward and sideways. They can cross water, mud, sand and level ground, which means that they are able to fly from shore to shore without having to use special ports. On a single trip, a big air-cushioned vehicle can take up to 400 passengers and 60 cars.

This ACV is a BHC AP. 1-88. It can carry 101 passengers with a top speed of 92 km/h (56mph).

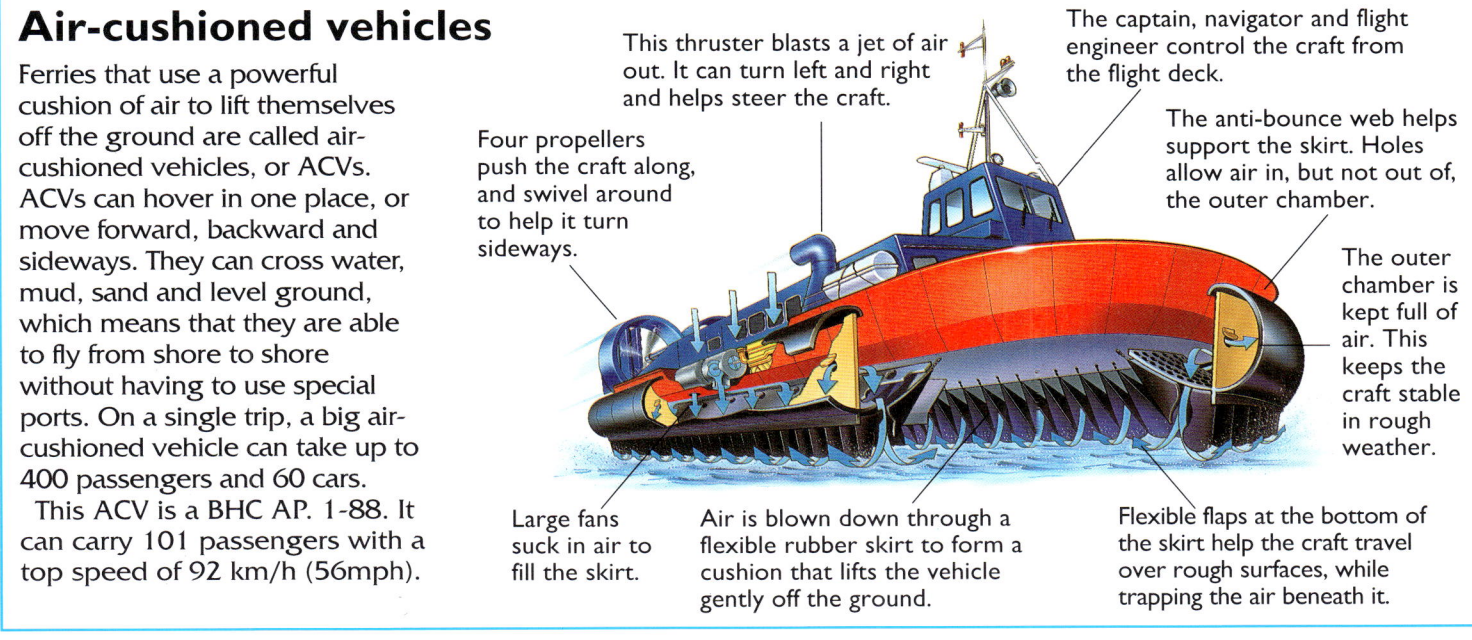

This thruster blasts a jet of air out. It can turn left and right and helps steer the craft.

Four propellers push the craft along, and swivel around to help it turn sideways.

The captain, navigator and flight engineer control the craft from the flight deck.

The anti-bounce web helps support the skirt. Holes allow air in, but not out of, the outer chamber.

The outer chamber is kept full of air. This keeps the craft stable in rough weather.

Large fans suck in air to fill the skirt.

Air is blown down through a flexible rubber skirt to form a cushion that lifts the vehicle gently off the ground.

Flexible flaps at the bottom of the skirt help the craft travel over rough surfaces, while trapping the air beneath it.

79

Engines

The engines of modern vessels range in size, from tiny outboards a child can lift, to diesels the size of a room and gas turbines as powerful as the jets on an airliner.

Some marine engines run on a mixture of petrol and oil, others use diesel fuel. Most are designed to run at a steady speed for long periods of time while the vessel cruises. Stop-start driving (such as a car faces in traffic) is unusual. Engines mostly drive propellers that range from the size of your hand, to about 7m (23ft) across in the case of supertankers. On some fast ships, engines drive water jets instead, because they reach much higher speeds with less wear and tear.

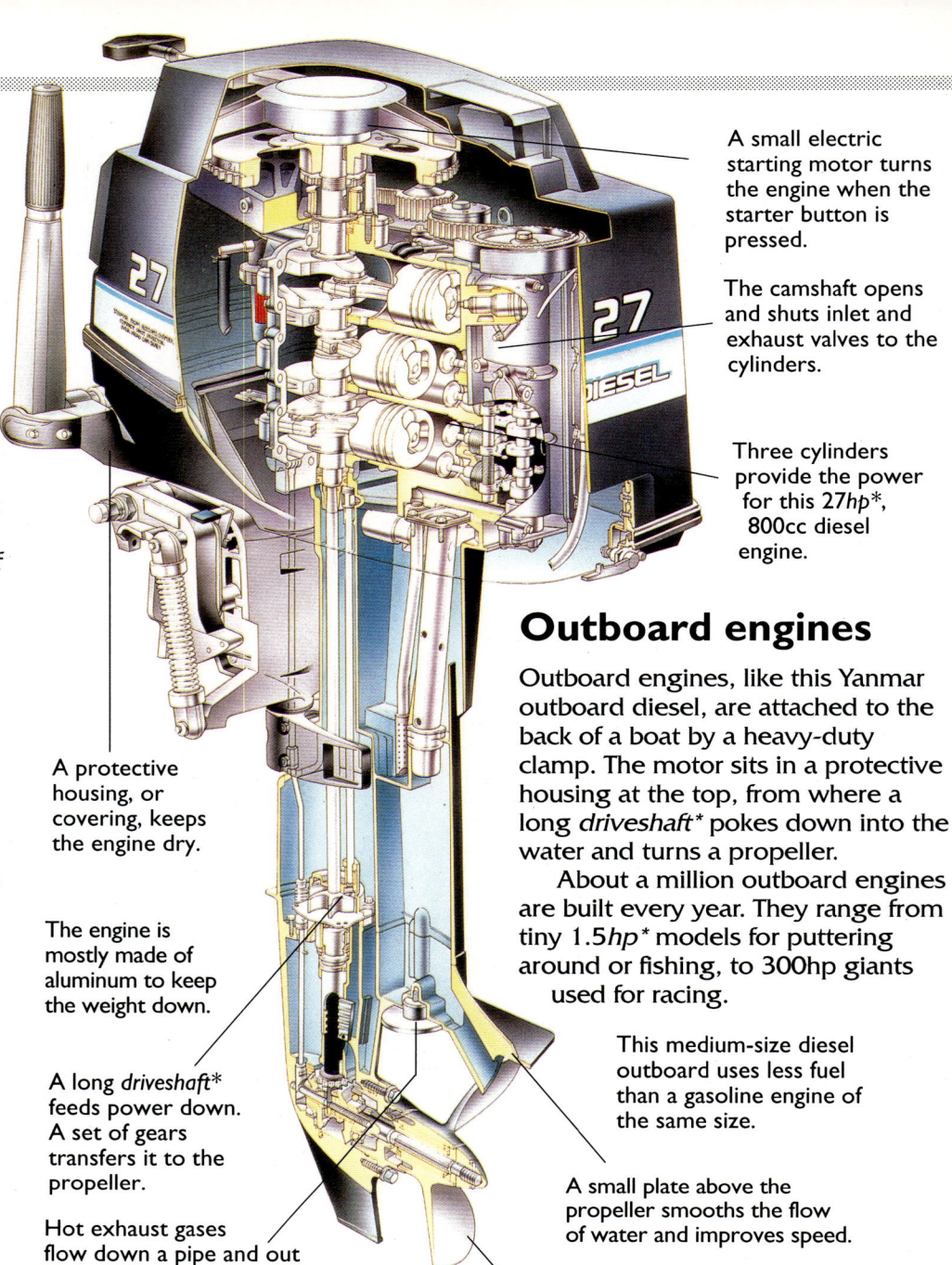

Electric power

This Yamaha electric drive outboard engine can propel a small boat with hardly any noise. It weighs less than 9 kg (20lbs), runs on a 12- volt car battery and produces 1/3 hp* of thrust. It's perfect for watching wildlife, or finding fish without disturbing them.

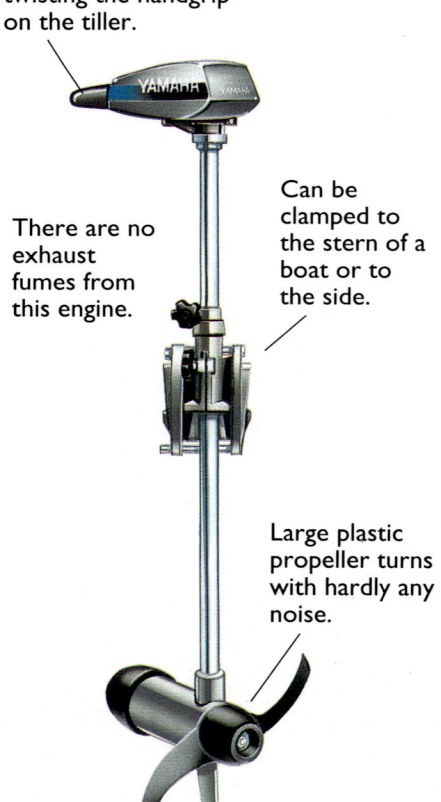

Speed is controlled by twisting the handgrip on the tiller.

There are no exhaust fumes from this engine.

Can be clamped to the stern of a boat or to the side.

Large plastic propeller turns with hardly any noise.

A small electric starting motor turns the engine when the starter button is pressed.

The camshaft opens and shuts inlet and exhaust valves to the cylinders.

Three cylinders provide the power for this 27hp*, 800cc diesel engine.

Outboard engines

Outboard engines, like this Yanmar outboard diesel, are attached to the back of a boat by a heavy-duty clamp. The motor sits in a protective housing at the top, from where a long *driveshaft** pokes down into the water and turns a propeller.

About a million outboard engines are built every year. They range from tiny 1.5hp* models for puttering around or fishing, to 300hp giants used for racing.

A protective housing, or covering, keeps the engine dry.

The engine is mostly made of aluminum to keep the weight down.

A long *driveshaft** feeds power down. A set of gears transfers it to the propeller.

Hot exhaust gases flow down a pipe and out into the water, just above the propeller.

This medium-size diesel outboard uses less fuel than a gasoline engine of the same size.

A small plate above the propeller smooths the flow of water and improves speed.

Propeller

How do propellers push?

As they spin, propeller blades force water to rush backward. The flow creates a strong thrust that shoves a boat forward. Big, slow-turning propellers have the strongest thrust of all.

Because propeller blades are curved, like those of a plane engine, (see page 43) water flows faster over the front of the blade than the back. This creates a second strong force, of *suction**, that also pulls the propeller forward as it turns.

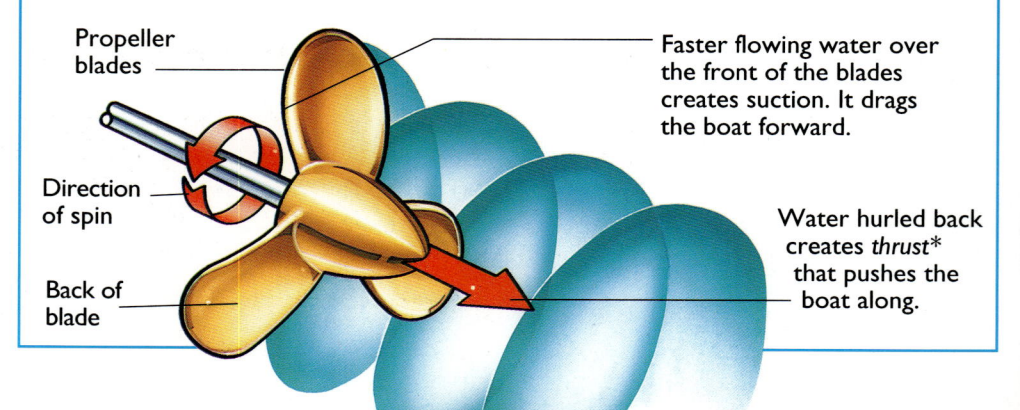

Propeller blades

Direction of spin

Back of blade

Faster flowing water over the front of the blades creates suction. It drags the boat forward.

Water hurled back creates *thrust** that pushes the boat along.

80

Inboard engines

Inboard diesel engines are the workhorses of the sea. They are used by yachts, fishing boats and all kinds of work boats, from tugs to supertankers. They are tough, strong, thrifty with fuel and can run for hours with next to no servicing.

This small diesel engine is used by sailing yachts to motor in and out of port, or to cruise when the sails are down. It runs quietly and smoothly and is extremely reliable.

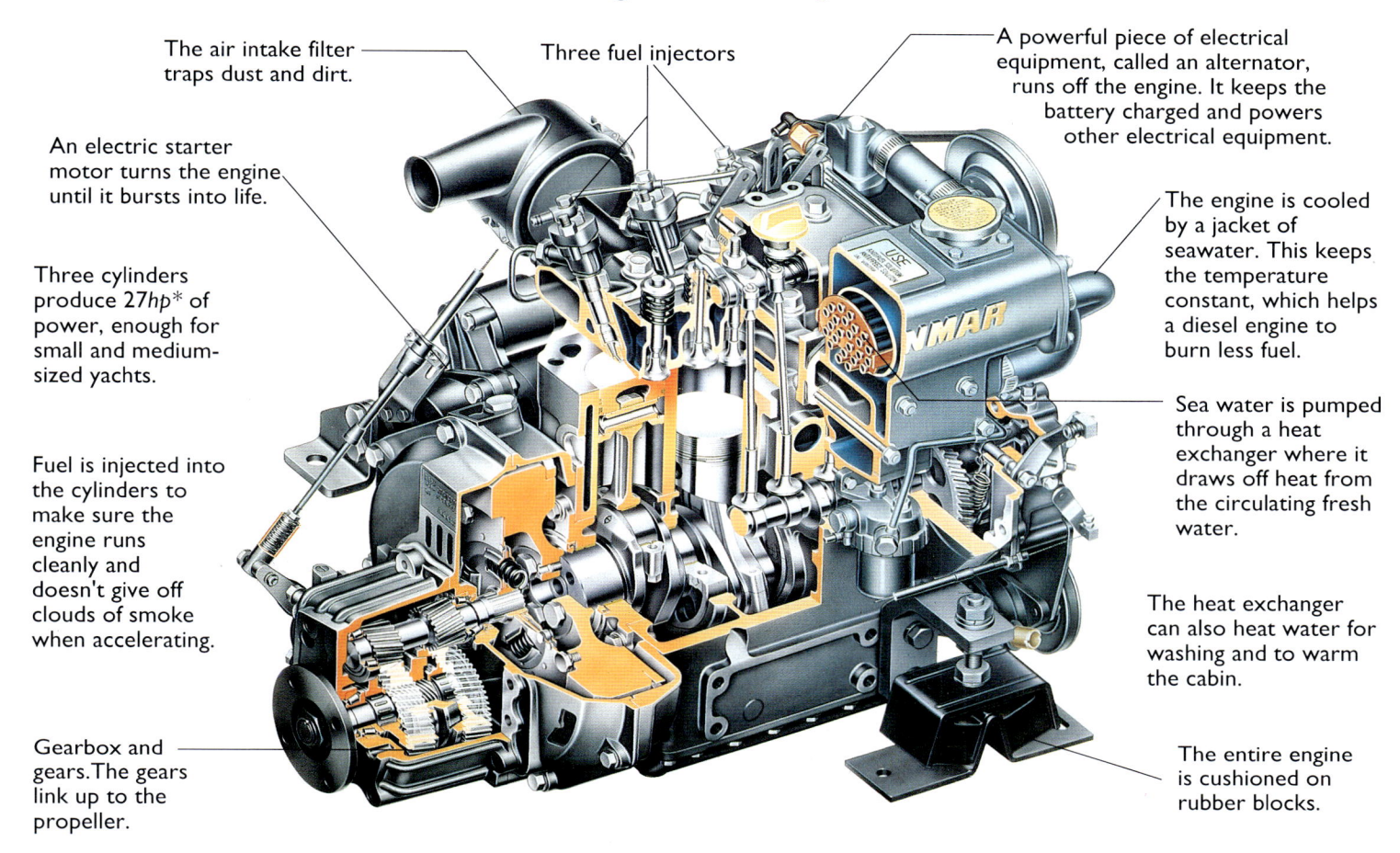

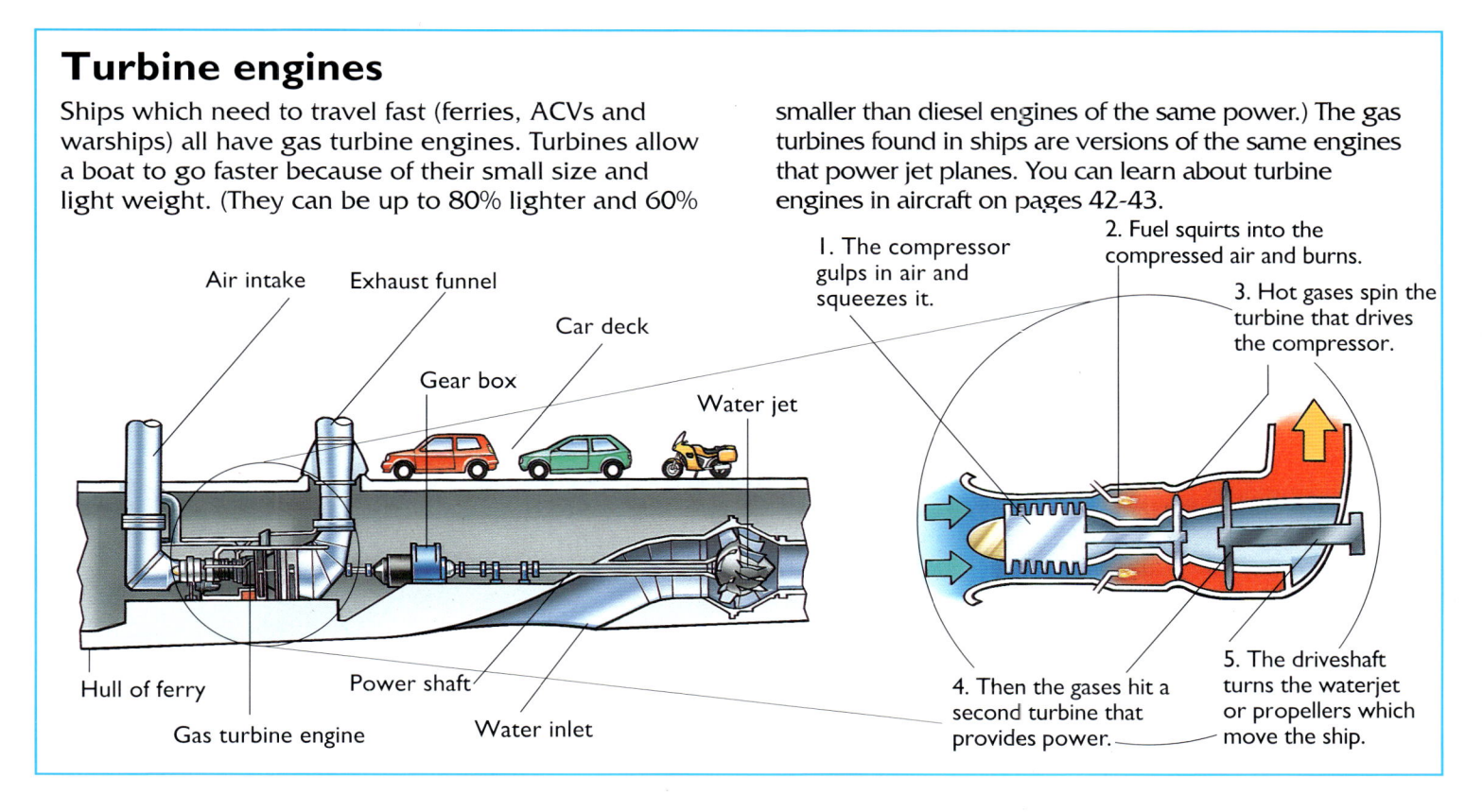

- The air intake filter traps dust and dirt.
- Three fuel injectors
- A powerful piece of electrical equipment, called an alternator, runs off the engine. It keeps the battery charged and powers other electrical equipment.
- An electric starter motor turns the engine until it bursts into life.
- Three cylinders produce 27hp* of power, enough for small and medium-sized yachts.
- Fuel is injected into the cylinders to make sure the engine runs cleanly and doesn't give off clouds of smoke when accelerating.
- The engine is cooled by a jacket of seawater. This keeps the temperature constant, which helps a diesel engine to burn less fuel.
- Sea water is pumped through a heat exchanger where it draws off heat from the circulating fresh water.
- The heat exchanger can also heat water for washing and to warm the cabin.
- Gearbox and gears. The gears link up to the propeller.
- The entire engine is cushioned on rubber blocks.

Turbine engines

Ships which need to travel fast (ferries, ACVs and warships) all have gas turbine engines. Turbines allow a boat to go faster because of their small size and light weight. (They can be up to 80% lighter and 60% smaller than diesel engines of the same power.) The gas turbines found in ships are versions of the same engines that power jet planes. You can learn about turbine engines in aircraft on pages 42-43.

- Air intake
- Exhaust funnel
- Car deck
- Gear box
- Water jet
- Hull of ferry
- Power shaft
- Gas turbine engine
- Water inlet

1. The compressor gulps in air and squeezes it.
2. Fuel squirts into the compressed air and burns.
3. Hot gases spin the turbine that drives the compressor.
4. Then the gases hit a second turbine that provides power.
5. The driveshaft turns the waterjet or propellers which move the ship.

Lifeboats

Lifeboats are some of the toughest working boats in the world. They are built to go to sea in really foul weather, and to work in waters strewn with rocks and sandbanks. Bucking high winds and breaking seas, they steer right alongside stricken ships to pass towlines or lift off crew.

Trent Class lifeboat

Some of the busiest shipping lanes in the world are next to Britain and Ireland. Apart from oil tankers, cargo ships and ferries, thousands of small fishing boats and yachts go to sea every day. Each year, lifeboats are called out over 6000 times (that's some 16 launches a day), saving more than 1,600 lives. The boats, like the Trent class lifeboat shown here, are all run by the RNLI (the Royal National Lifeboat Institution).

The Trent Lifeboat
Length: 14.26m (46ft 9in)
Weight: 26.5 tonnes (26 tons)
Speed: 25 knots
Range: 400km (250 miles)
Crew: 6
Engines: Two 800hp diesels

This RNLI Trent class lifeboat covers one stretch of coast and up to 80 km (50 miles) out to sea.

The boat has antenna for several radios and for radar. It also uses satellite navigation and a signal tracker that homes in on radio messages from ships in distress.

Powerful searchlights to work in poor visibility and at night

The coxswain, who is in charge, may command the lifeboat from the upper platform. Here he can see all around while steering and talking on the radio.

Boxes holding emergency life rafts

Bollard to fasten rope when towing small boats to safety

Life rings

Rear flaps, called trim tabs, change the angle of the bow to suit sea conditions.

The hull has three water inlets. Water flows into two of them to cool the big engines. The third feeds a fire pump in the engine room.

The boat is completely watertight. All air inlets and outlets have seals to keep out water.

Twin 800-*horsepower** diesel engines, each as powerful as eight small cars

The right way up!

Lifeboats are built of tough lightweight materials that are completely watertight. They will bob upright almost at once if a wave ever knocks them flat. As they flip over, their engines automatically slow down. Then, as soon as the boats right themselves, the coxswain simply opens the throttle and continues on his way.

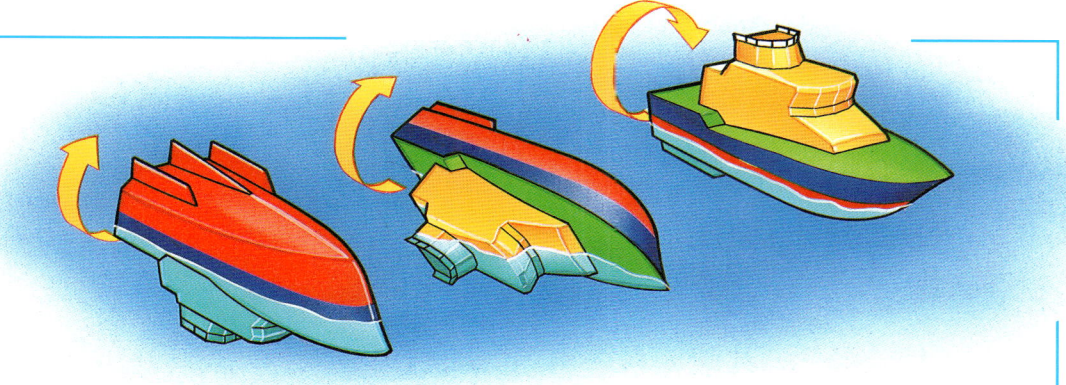

1. All new lifeboats are capsize-tested for safety. A crane tilts the boat into the water until it is lying completely upside down.

2. As the crane lines are dropped, the boat quickly turns itself the right way up. Water pours from the upper decks as it rights itself.

3. The heavy engines are so low down, and there is so much air in the cabin and hull, that the lifeboat flips upright without needing help.

Charts and radar screens

Front hatch and ladder to survivors' cabin

In the survivors' cabin is a galley (kitchen).

The hull is made from a sandwich of tough plastics and very light foam. There are no steel beams anywhere to add extra weight.

A crew of six rides in the wheelhouse (a cabin on the bridge). There are ten seats in the cabin below for survivors.

Crewmen wear helmets and sit harnessed in highback seats that stop them from flying about as waves smash into the boat.

Fuel tanks are filled with foam, just like tanks in some racing cars. It stops fuel from sloshing around as the boat rolls.

The sides of the deck are extra low to make it easier to snatch people from the sea.

Side keels

A lifeboat may work so close to shore that a really big wave can make it touch bottom. To protect the propellers, it has a pair of very deep keels on either side of the main keel. These reach further down than the propeller blades and so will touch bottom first. They also keep the boat upright at low tide if it gets stranded away from base.

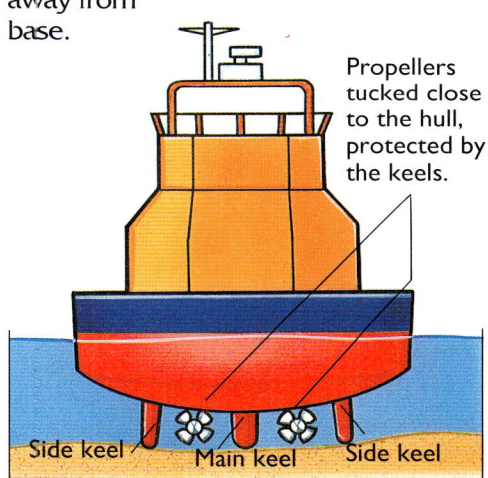

Propellers tucked close to the hull, protected by the keels.

Side keel Main keel Side keel

A quick launch

Some lifeboats are moored afloat, but others are stationed in boathouses. In order to put to sea, they use a greased slope called a slipway. First their engines are started. Then, once the single holding wire is released, the lifeboat slides down the slipway and gathers speed. It hits the water at almost nine *knots*.

A groove in the slipway guides the keel.

The side keels keep the lifeboat upright.

83

Racing boats

Racing powerboats are designed to rise out of the water and skim the surface at high speed. There are three basic types: monohulls, catamarans and hydroplanes.

A monohull is another name for a single-hulled boat. (Most non-racing boats belong to this category.) A catamaran has two narrow hulls, one on each side of the driver. The hulls are set wide apart to make the boat stable. A hydroplane is a half-breed. It has two hulls in front, while the back half narrows into a single hull.

Hull shapes

Monohulls have a flaring V-shape that helps the front section of the hull to rise out of the water at high speed.

Only the rear hulls of catamarans (and their propeller and rudder) stay wet at race speeds, making them the fastest class of racer.

The twin front hulls of hydroplanes create *lift**. At full speed they become airborne. Only the back hull stays in the water.

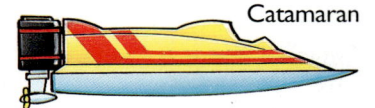

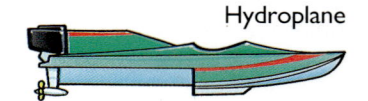

Monohull

Catamaran

Hydroplane

Ocean racer

Surfury was a monohull cruiser designed for offshore races in heavy seas and winds. It was built in 1965 and, over the next five years, carved out a reputation as one of the world's best racers in long distance events.

A third crew member tended the engines. He stood behind the drivers.

Surfury won the British Cowes-Torquay race in 1967 with an average speed of 85km/h (53 mph).

Two big Daytona engines, one behind the other, provided 1050 *horsepower**.

Surfury was 11m (36ft) long. It was built from sheets of laminated cedar wood, pressed into shape.

Two drivers rode half-standing, supported by reclining seats that cushioned them from the battering of high-speed travel over waves.

Part of the cabin roof was replaced with a tarpaulin to save weight.

A tiny galley (kitchen) enabled the crew to prepare meals.

One shaft and propeller, instead of two, cut down *drag** and made the boat much faster.

The front engine's hot exhaust was piped over the side. The rear engine's exhaust was vented through the stern.

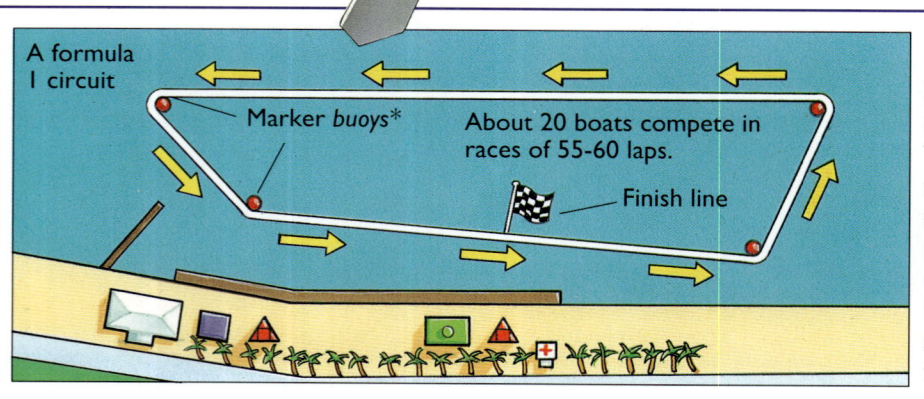

A formula 1 circuit

Marker *buoys**

About 20 boats compete in races of 55-60 laps.

Finish line

Formula 1 racing

Formula 1 boats are small, streamlined catamarans with a huge outboard engine. Their hulls are built from synthetic materials that are light but immensely strong, to withstand pounding at top speeds. Like Formula 1 cars, these boats compete all over the world. There are usually about 12 events a year, held in sheltered waters where boats can reach top speeds of 260 km/h (165 mph).

Plane speed

Racing boats are built with specially shaped bottoms, so their hulls can plane (or skim) across the surface, rather than carve a path through the water. This increases their speed enormously, because the engines avoid wasting power by having to push aside a heavy weight of water.

A Victory boat is a type of big catamaran that has won many offshore races in recent years.

Circuit racing ranges from Formula 1 boats, to little J250 craft, like this, that children of nine and up can race.

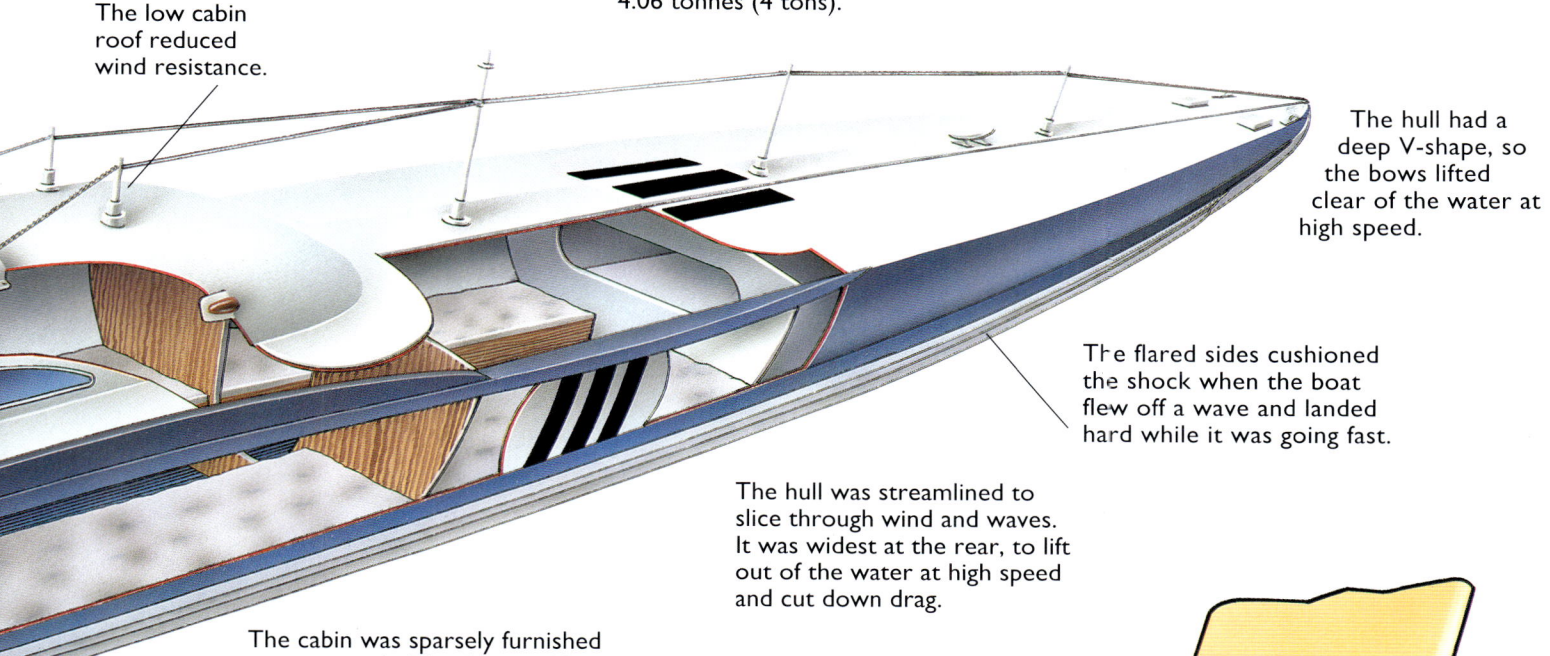

The low cabin roof reduced wind resistance.

The racing weight was 4.06 tonnes (4 tons).

The hull had a deep V-shape, so the bows lifted clear of the water at high speed.

The flared sides cushioned the shock when the boat flew off a wave and landed hard while it was going fast.

The hull was streamlined to slice through wind and waves. It was widest at the rear, to lift out of the water at high speed and cut down drag.

The cabin was sparsely furnished to save weight. The dining table doubled as the door to the toilet.

Hydrofoils

The speed of any vessel in water is limited by a force called *drag**, created by the friction between the boat and the water. This means that ordinary boats cannot travel much faster than 35 km/h (20mph). So, with a given size of boat and engine, the easiest way to boost a boat's speed is by lifting the hull out of the water altogether. One of the best ways of doing this is with hydrofoils.

Hydrofoils are flat struts fixed to the hull below water. They are shaped like the wings of a plane (see page 35). As a boat gathers speed, water flows faster over the curved upper side of a hydrofoil than the flat surface beneath it. Low pressure forms above the foil and, as with a plane wing, creates *lift**. The strut rises up. As the hull lifts out of the water, drag decreases. Now, running with the same power, the boat swiftly picks up speed. Big passenger hydrofoils can accelerate up to more than 90km/h (almost 60 mph).

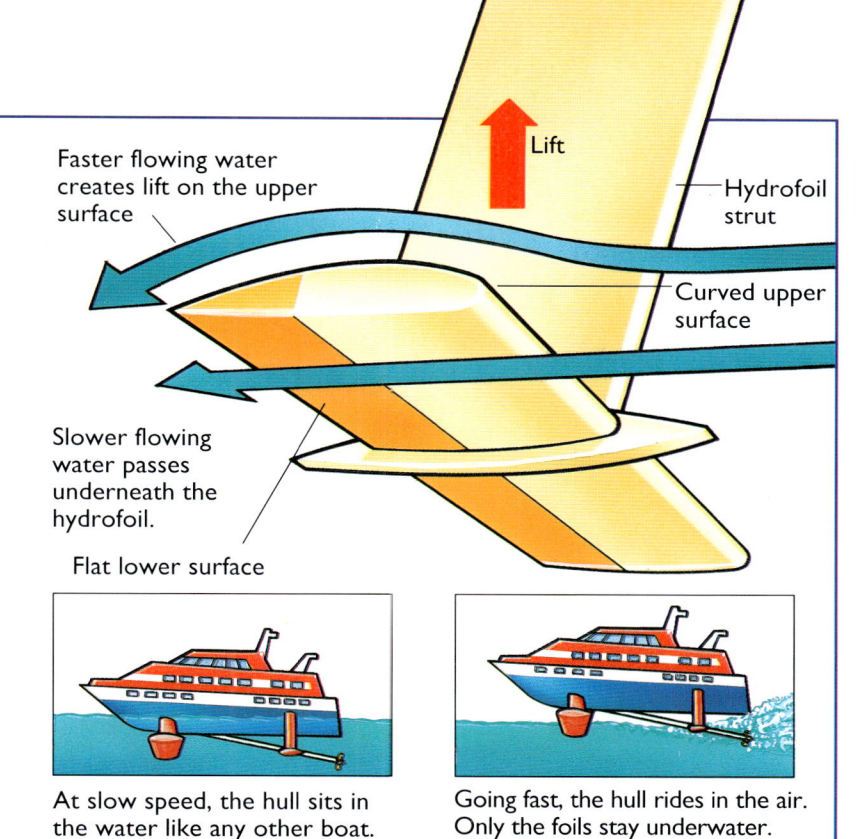

Faster flowing water creates lift on the upper surface

Lift

Hydrofoil strut

Curved upper surface

Slower flowing water passes underneath the hydrofoil.

Flat lower surface

At slow speed, the hull sits in the water like any other boat.

Going fast, the hull rides in the air. Only the foils stay underwater.

85

Container ships

Modern cargo ships are huge and expensive to build. So they are designed to spend as little time as possible resting in port. To make loading quicker, nearly all freight goes on board in gigantic, prepacked metal boxes, called containers. Other freight is designed to be driven, or towed by trailers, on and off - just like cars on a ferry. This kind of freight is known as roll-on roll-off (or ro-ro). It can include anything from railway carriages to helicopters and earth-moving equipment.

The *Atlantic Companion* (below) is one of five 53,000-tonne (52,000-ton) G3 models owned by the Atlantic Container Line. They are among the largest combination container/ro-ro ships afloat.

The *Atlantic Companion* carries containers and ro-ro cargo from the USA to Europe. Each crossing takes six to eight days.

Atlantic Companion	
Length:	292m (958ft)
Width:	32m (105ft)
Size:	53,000 tonnes (52,000 tons)
Draft:	11m (36ft)
Engine:	27,500 hp* diesel

Dining room and day room where the crew can take breaks

Indoor swimming pool and sauna

Five levels of cars can park in the upper garage.

Library and TV/video room

The whole ship is controlled from a room called a wheelhouse that runs the full width of the ship. The ship is steered by computer, while at sea.

There are two 50-person lifeboats - one on each side of the ship.

Refrigerated containers

A wide ramp lowers from the back of the ship. Two lanes of traffic at a time can use the ramp - one loading and the other unloading.

A single main propeller can drive a loaded ship at a cruising speed of 18 *knots**.

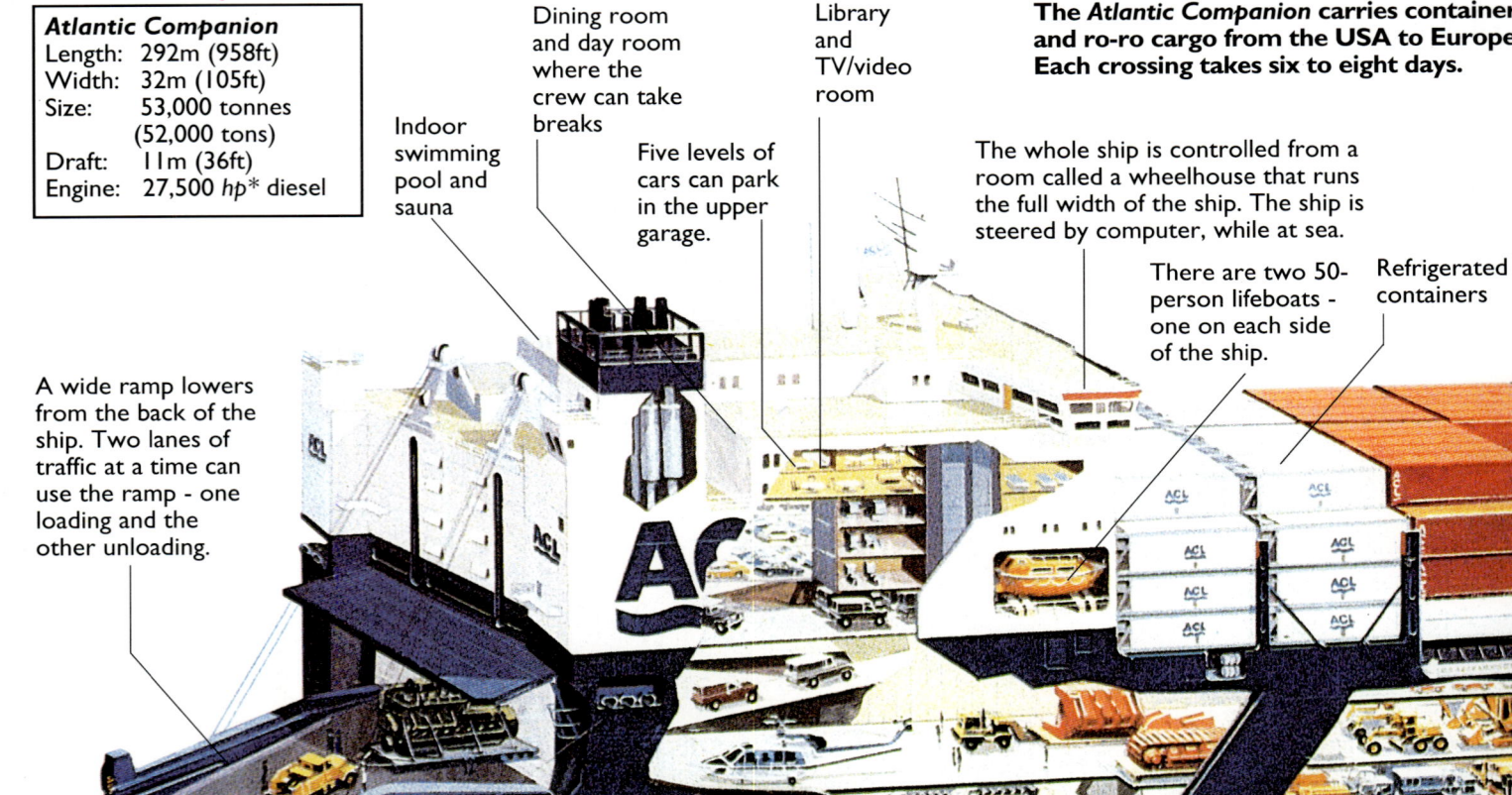

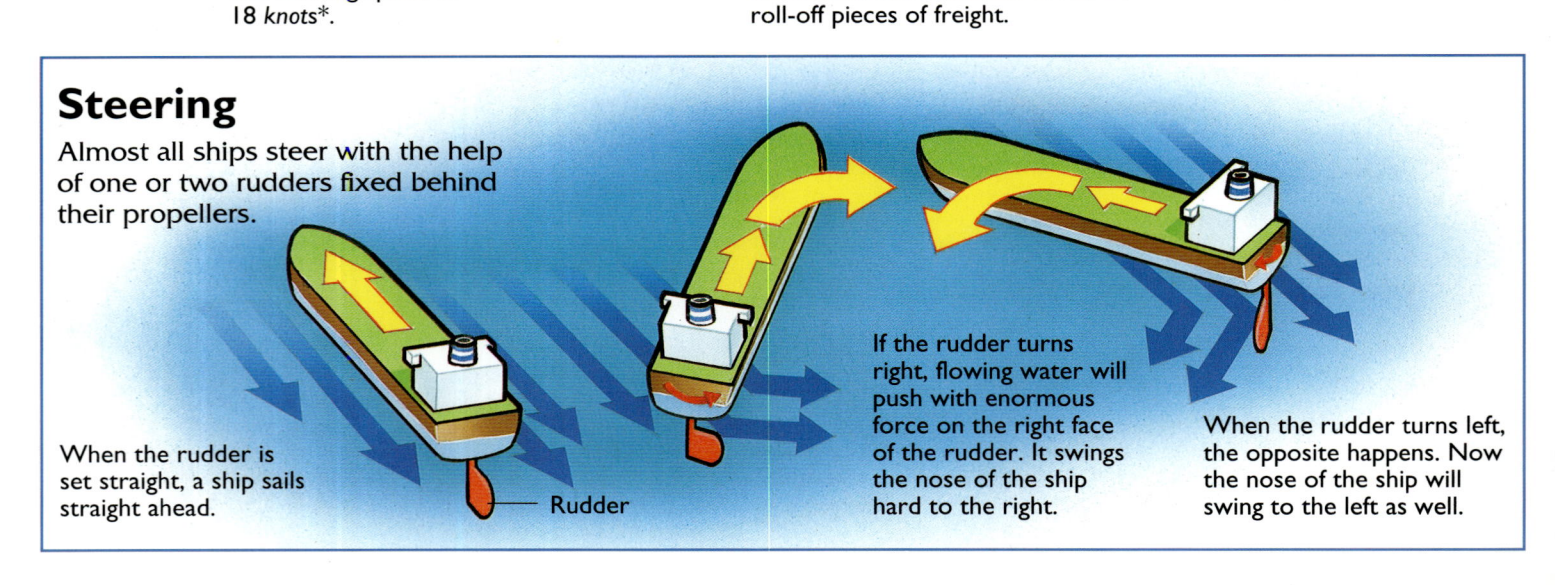

Stern *thruster** for docking (see box opposite)

Fully loaded, the ship carries 1045 cars, over 1900 containers, 175 refrigerated containers, and hundreds of roll-on roll-off pieces of freight.

A giant six-cylinder diesel engine drives the ship.

Steering

Almost all ships steer with the help of one or two rudders fixed behind their propellers.

When the rudder is set straight, a ship sails straight ahead.

Rudder

If the rudder turns right, flowing water will push with enormous force on the right face of the rudder. It swings the nose of the ship hard to the right.

When the rudder turns left, the opposite happens. Now the nose of the ship will swing to the left as well.

Container ports

Some of the busiest ports in the world, such as Hong Kong and Singapore, each handle over 10 million containers a year.

After a ship docks, giant cranes, able to lift 50 tonnes (49 tons) at a time, roll into position to unload the containers.

The cranes lower the containers onto special trucks which carry them to huge parking lots.

Lifting trucks stack them in rows.

The containers are loaded onto railway cars or trailer trucks and taken to their final destination.

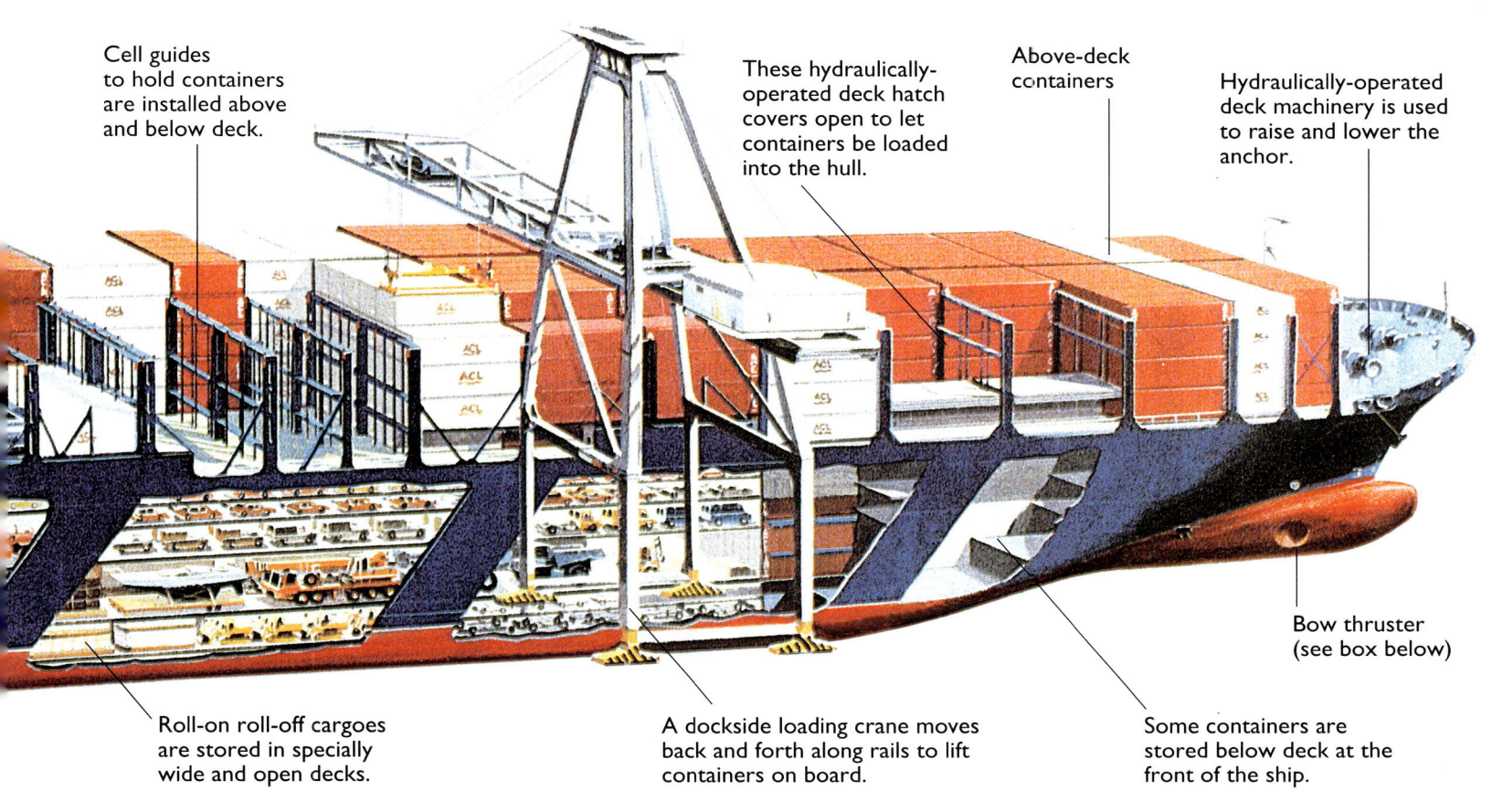

Cell guides to hold containers are installed above and below deck.

These hydraulically-operated deck hatch covers open to let containers be loaded into the hull.

Above-deck containers

Hydraulically-operated deck machinery is used to raise and lower the anchor.

Roll-on roll-off cargoes are stored in specially wide and open decks.

A dockside loading crane moves back and forth along rails to lift containers on board.

Some containers are stored below deck at the front of the ship.

Bow thruster (see box below)

Thrusters

All G3 ships can dock without help from tugboats. They use big propellers, called thrusters, attached to the bow and stern to create a sideways blast of water.

Run together, in the same direction, the thrusters slowly nudge the ship sideways as it docks. Pushing on opposite sides, they turn the ship around within its own length.

Thrusters work by pushing water from one side of the ship to the other through a large tunnel in the hull.

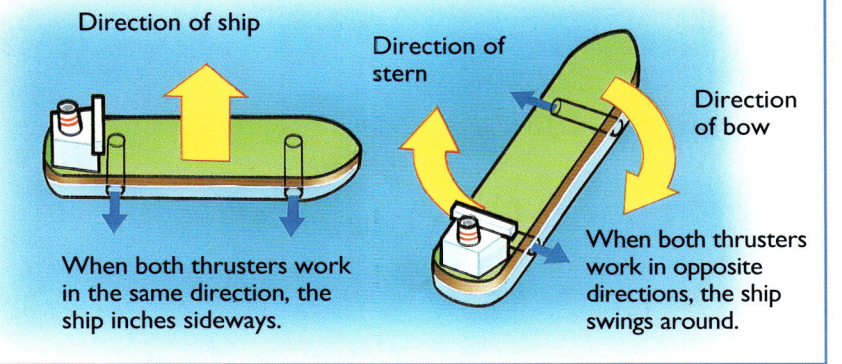

Direction of ship

Direction of stern

Direction of bow

When both thrusters work in the same direction, the ship inches sideways.

When both thrusters work in opposite directions, the ship swings around.

Tugs

Ocean-going ships are so big that they are difficult to steer in enclosed waters. This means they have trouble sailing in and out of port. That's where tugs come in.

Tugs are stubby little boats that stop, start and turn with ease. They handle so well they can work in even the tightest spaces, alongside piers, or in closed-off sections of canals or rivers called locks.

Tucked into a tug's hull is an incredibly powerful engine that drives a huge propeller. This provides the power to tow cargo ships and oil tankers well over a hundred times as heavy as the tug.

A 20 year-old tug
Length:	32m (105ft)
Draught:	4.7m (15ft)
Engines:	Twin diesels
Top speed:	About 12 knots
Crew:	Up to 12
Pulling power:	around 40 to 50 tonnes (39-49 tons)

All-purpose tugs

A general purpose tug doesn't work only with ships. It may also tow barges and dredgers, fight fires (see right) with its pumps and hoses, or mop up oil spills left by tankers. Tugs are sometimes hired to carry crews and other passengers from ship to shore, or as rescue craft to help ships that get into trouble.

This is a deep-sea tug, which is longer and heavier than the ones used in ports. Their high bows are built to shrug off ocean waves, while their huge engines can tow anything from oil rigs to crippled supertankers.

Exhaust funnel for diesel engine

Steel bars to keep tow lines from snagging on the deck

Tow ropes are 15cm (6in) thick and able to take a strain of over 100 tonnes (98 tons).

Capstan winch for reeling in heavy tow ropes

Pumps to feed water or foam to fire guns

Rudder

Two diesel engines can produce 3,000hp - a huge amount of power for such a small boat.

Propeller power

Many new tugs have two sets of special high-powered propellers called Voith-Schneider propellers. They look a bit like egg-beaters, and sit mid-hull instead of at the stern. Unlike ordinary propellers, they can thrust in any direction. This lets tugs tow at full power in whatever direction the captain wants to go.

Each blade looks like a stubby little wing.

When the blades swing at an angle during part of each turn, the ship starts to move - in this case forward.

When the ship is at rest, the blades turn without an angle and so create no thrust.

A deep keel helps to steer.

Struts around the blades boost the thrust of the propellers.

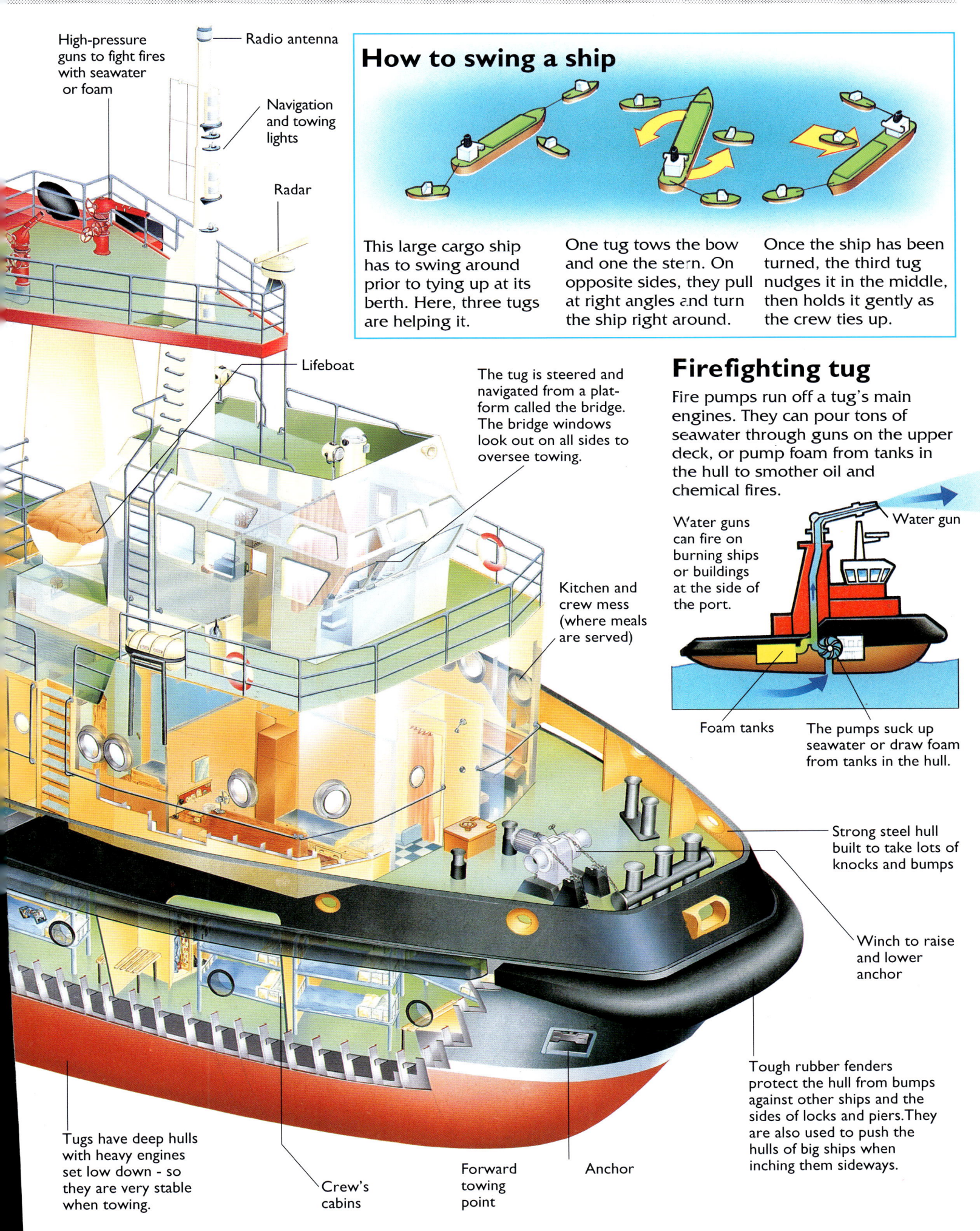

High-pressure guns to fight fires with seawater or foam

Radio antenna

Navigation and towing lights

Radar

Lifeboat

The tug is steered and navigated from a platform called the bridge. The bridge windows look out on all sides to oversee towing.

Kitchen and crew mess (where meals are served)

Tugs have deep hulls with heavy engines set low down - so they are very stable when towing.

Crew's cabins

Forward towing point

Anchor

Strong steel hull built to take lots of knocks and bumps

Winch to raise and lower anchor

Tough rubber fenders protect the hull from bumps against other ships and the sides of locks and piers. They are also used to push the hulls of big ships when inching them sideways.

How to swing a ship

This large cargo ship has to swing around prior to tying up at its berth. Here, three tugs are helping it.

One tug tows the bow and one the stern. On opposite sides, they pull at right angles and turn the ship right around.

Once the ship has been turned, the third tug nudges it in the middle, then holds it gently as the crew ties up.

Firefighting tug

Fire pumps run off a tug's main engines. They can pour tons of seawater through guns on the upper deck, or pump foam from tanks in the hull to smother oil and chemical fires.

Water guns can fire on burning ships or buildings at the side of the port.

Water gun

Foam tanks

The pumps suck up seawater or draw foam from tanks in the hull.

89

Cruise ships

Few people cross oceans by ship any more. Planes are much faster and more convenient. Although ocean liners have long gone, their place has now been taken by cruise liners. These ships are designed specifically for pleasure trips, usually calling at a number of different ports.

The Sun Princess

The *Sun Princess*, launched in 1995, is one of the newest big ships to join the fleet that sails the Caribbean in winter and the Alaska Coast in summer. Cruise ships take over 4.5 million people on trips every year.

At 78,250 tonnes (77,000 tons) and 261m (856ft), the *Sun Princess* is the biggest cruise ship afloat.

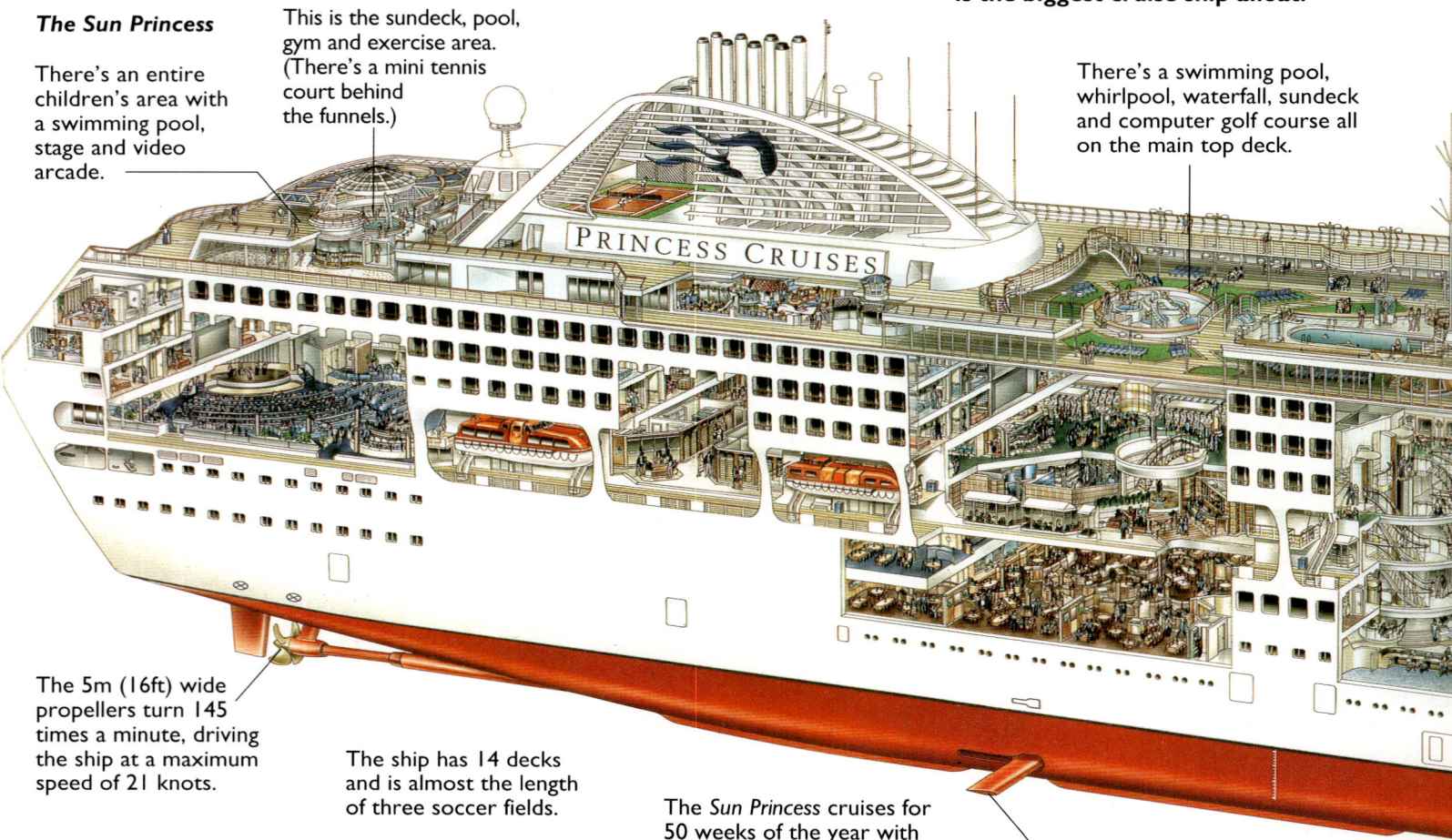

The Sun Princess

There's an entire children's area with a swimming pool, stage and video arcade.

This is the sundeck, pool, gym and exercise area. (There's a mini tennis court behind the funnels.)

There's a swimming pool, whirlpool, waterfall, sundeck and computer golf course all on the main top deck.

The 5m (16ft) wide propellers turn 145 times a minute, driving the ship at a maximum speed of 21 knots.

The ship has 14 decks and is almost the length of three soccer fields.

The *Sun Princess* cruises for 50 weeks of the year with passengers on board.

A stabilizer fin juts out on each side of the ship.

Silent night

The cruising speed of the *Sun Princess* is 21 knots (a good 10 knots slower than older passenger liners like the *QE2*). Its two propellers each have six curved blades, which draw water past the hull with little turbulence. Each propeller is driven by an electric motor, mounted on rubber to cut down noise.

How stabilizers work

Two stubby wings called stabilizers poke from the hull of the ship below the water line. They smooth out the rolling motion of the waves.

These wings waggle back and forth all the time, controlled by computers in the ship that sense exactly what the waves are doing.

As waves wash past, they set up a side-to-side rolling motion.

The stabilizer helps to keep the ship in a more upright position.

The stabilizers fold into a bay in the sides when the sea is calm. As the sea gets rougher, a single stabilizer is brought into use.

It acts like the wing of a plane, lifting one side of the ship in the opposite direction to the rolling motion of the waves.

The two opposing rolling motions cancel each other out. This means the ship continues in a steadier, more upright manner.

A floating hotel

Part resort, part luxury hotel, cruise ships are amazingly comfortable. In the best suites, passengers can enjoy marble bathrooms, whirlpool tubs, TVs and private bars. Over 400 cabins come with outside balconies overlooking the sea.

In the heart of the ship is a four-floor high lobby where people can glide up and down in glass elevators.

Sea fare

With four restaurants and cafés, and five bars scattered around the ship, passengers can eat and drink just about any time, day or night.
A typical shopping list for a seven-day cruise might include the following groceries:

10,886 kg (24,000 lbs) of meat
2994 kg (6600 lbs) of fish
726 kg (1600 lbs) of fresh shrimp
4082 kg (9000 lbs) of potatoes
15,876 kg (35,000 lbs) of fruit
Enough coffee to brew 8865 ltrs (1,950 gallons)

Radar antennae, radio equipment and satellite links for phones, faxes, computers and TV.

The ship can take as many as 2,022 passengers at a time, with a crew of 920.

The bridge, the area from which the captain and his officers run the ship

Computers guide and steer the ship. They can stop her outside a port and hold her steady there without ever needing to drop the anchor.

The bulb shaped bow parts the waves to enable the ship to slip through the sea with less effort.

The ship has a large auditorium in the front for shows and concerts.

Bulkheads

The hull of a cruise ship, below the water line, is divided into compartments by watertight walls called bulkheads. These are designed so that if water gets into one compartment it doesn't spread through the ship.
  Above this lies the watertight bulkhead deck. No water can rise above it, even if the lower hull is entirely flooded. So, even if a ship is sinking, bulkheads stop it from capsizing (rolling over) due to water rushing to one side. This gives people more time to escape.

All watertight bulkhead doors can be shut by remote control from the bridge.

Bulkhead deck

Bulkheads

The compartments below the bulkhead deck hold the engines, air-conditioning, supplies, laundries, and cabins for the crew.

The ship will still stay afloat, even if two compartments are flooded.

91

Submersibles

Submarines are built for military use. Navies use them to launch missiles or to sink surface ships.

Submersibles are something altogether different. They are small diving craft built for scientific research, archeology, or to work in oil and mineral exploration. Some map the seabed, others repair pipes and cables, and a few are used for rescue work. They can all dive far deeper (about four times as deep) than any military submarine.

Length:	7.6m (25ft)
Width:	2.4m (8ft)
Engines:	6 small electric thrusters*
Cruising speed:	1 knot*
Top speed:	1.5 knots
Max depth:	4000m (13,000ft)
Range:	8 kms (5 miles)

Two still cameras are fixed outside to take pictures while three video cameras record everything on film.

Alvin can dive to about 4,000m (15,000ft) - about as far as a person can walk in an hour.

The robot arm collects rocks and other samples from the ocean floor.

Alvin the submersible

One of the best-known submersibles is a little research vessel called *Alvin*. First launched in 1964, it has since made thousands of dives around the world. It was the first vessel to explore the wreck of the *Titanic*, and to discover belching vents of hot water at the bottom of the sea. These occur where cracks in the seabed have caused heat deep inside the Earth to raise the temperature of water seeping in by hundreds of degrees. In these isolated spots, scientists have found colonies of strange tube worms and shrimps not known anywhere else.

Three small portholes allow the crew to see out.

*Thrusters**

The seafloor where *Alvin* works is pitch black. Its powerful lights can only light up a small patch of seabed.

This is a remote-controlled robot called *Jason Jr*, that was used to explore the wreck of the *Titanic*. It was taken down in a cage, bolted to the front of *Alvin*, and steered by cable into the ship.

Video camera

Using special equipment, divers like this one from *Norbert* can work up to 250m (820ft) deep.

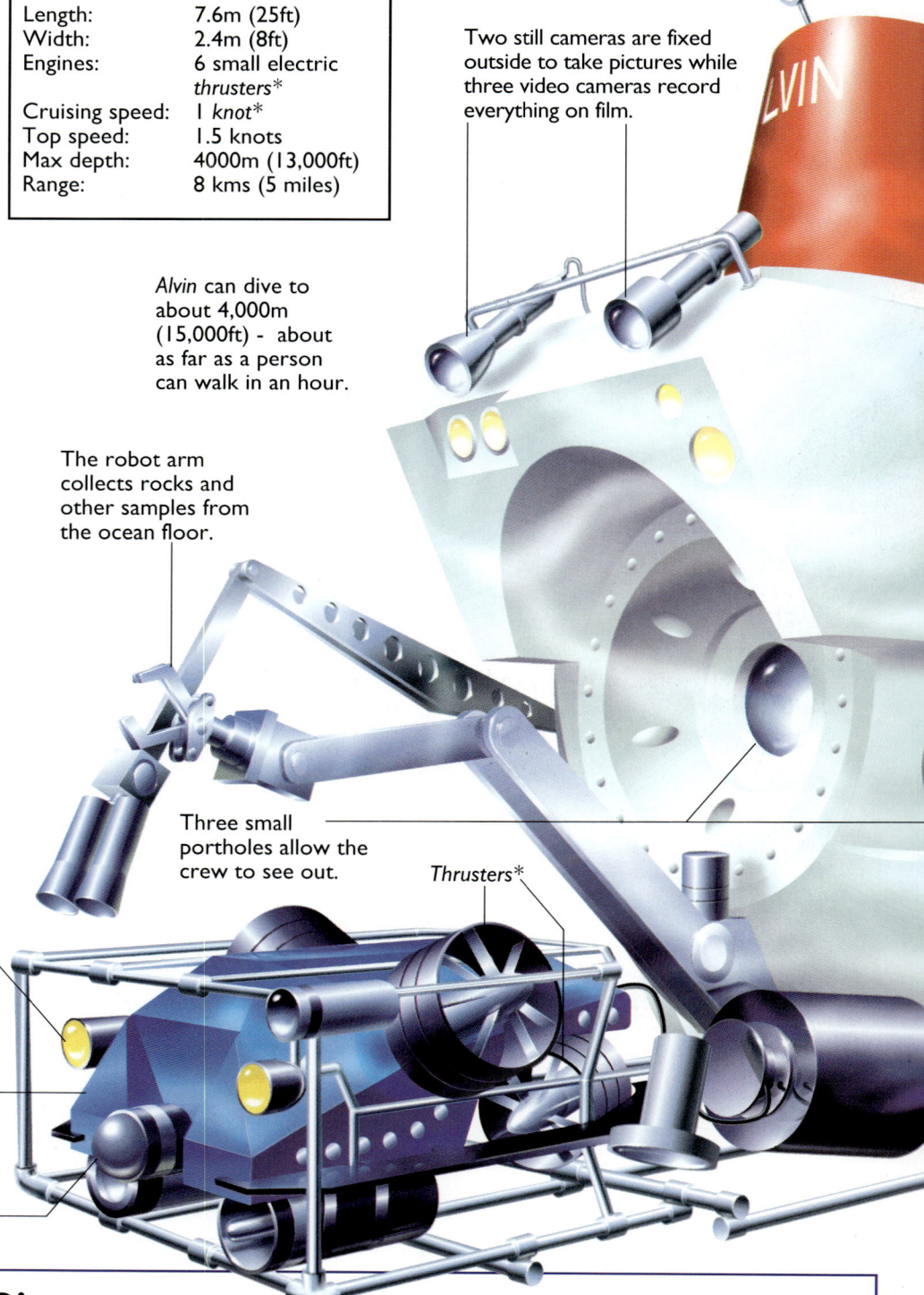

Big squeeze

The amount of air pressure at sea level is called 'one atmosphere'. Underwater, pressure builds up very quickly, as water is far heavier than air. Every 10m (33ft) farther down adds another 'atmosphere' of pressure.

The limit of *Alvin's* range is just under 4,000m (13,120ft). In the deepest parts of the ocean, almost 11,000m (36,000ft) down, pressure may be 1,000 times greater than at the surface. Here a submersible would crumple like an empty can.

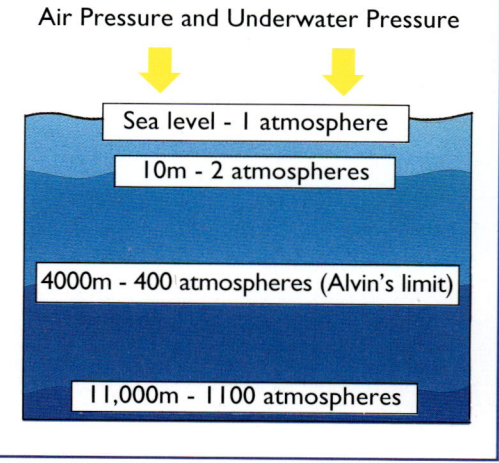

Air Pressure and Underwater Pressure

- Sea level - 1 atmosphere
- 10m - 2 atmospheres
- 4000m - 400 atmospheres (Alvin's limit)
- 11,000m - 1100 atmospheres

Military submarines

Military submarines are designed to operate in a shallow band of water, no more than about 200-300m (600-900ft) below the surface. They are used to attack surface ships, or hunt enemy submarines and fire missiles. Their hulls are strong, but very few can go deeper than 500m (1640ft).

This is a British Trafalgar Class military submarine at anchor. It can reach a top speed of 30 knots*.

The cramped cabin is a metal ball just over 2m (8ft) wide.

In an emergency, the passenger cabin can detach from the frame and float to the surface on its own.

Sets of tiny thrusters are used to drive and steer *Alvin*.

The passenger cabin is made of titanium. It is as hard as steel, but much stronger.

Rack of batteries to power *Alvin*

Air tanks and ballast tanks

The sub stays in contact with the surface by radio telephone.

Mini-subs

Mini submersibles are widely used, for example, in the oil business, to move divers, or to work at depths that are too dangerous for free-swimming humans. They are equipped with floodlights, cameras, robot arms, and a highly accurate navigation system so they can find their way about in pitch darkness.

As all these subs run on batteries, they can only stay under for a very short time (usually less than a day) before they surface and recharge.

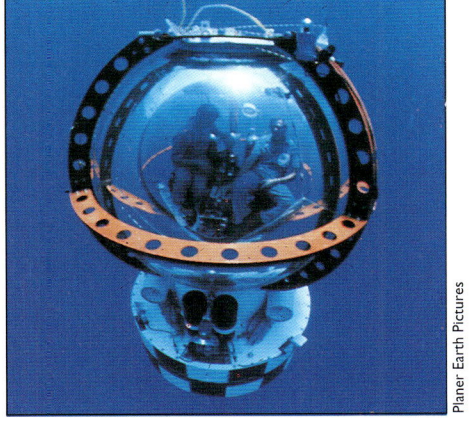

Nemo (Naval Experimental Manned Observatory) operates at a depth of 183m (600ft), with a crew of two.

How do subs dive?

Submersibles can only go up and down. They carry heavy lead weights as they dive that are dumped at the bottom when the vessels need to stop going down.

A submarine is different. As well as being able to travel on the surface and dive, it can also hover at whatever depth it wants to. This is possible because it has air tanks, called ballast tanks, all along the outside of its hull. They are open to the sea at the bottom and have vents at the top.

Floating on the surface, the ballast tanks are full of air and their vents are closed.

Air — Vents closed

To dive, the vents are opened to let water flood into the tanks. This makes the sub heavier and so it sinks.

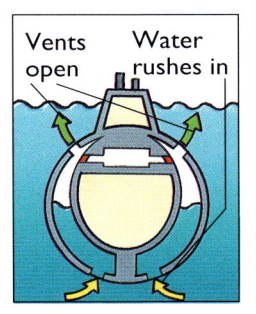

Vents open — Water rushes in

To hover or travel at the desired depth, the vents are closed, so the sub stops sinking. The tanks are full of water.

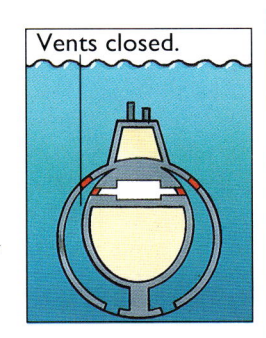

Vents closed.

To go up, high-pressure air is blown into the tanks, forcing out the water. The submarine rises to the surface.

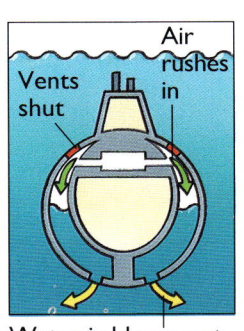

Vents shut — Air rushes in

Water is blown out.

93

The future

The look of boats has changed out of all recognition in the past 150 years. During this time, they have grown bigger, faster and much more comfortable. If change continues like this, it is likely that boats of the future will look very different from the way they do today. Since the main problem is the way water slows boats down, many of the newest ideas are concentrating on raising them out of the water, to make them go faster.

Air superliner

A group of Japanese companies are testing the idea of fast catamaran freighters, able to carry 1,000 tonnes (984 tons) at over 50 knots, with a range of 800 km (500 miles) or more. One model, the TSL-A, uses a cushion of air to lift much of its body out of the water in order to reach top speed. So far, only half-size models, like the *Hisho* (below), have been tested at sea, but they have been a great success.

Hisho is a 70m (230ft) half-size model of the TSL-A. It is driven by two 16,000hp engines, linked to water jets.

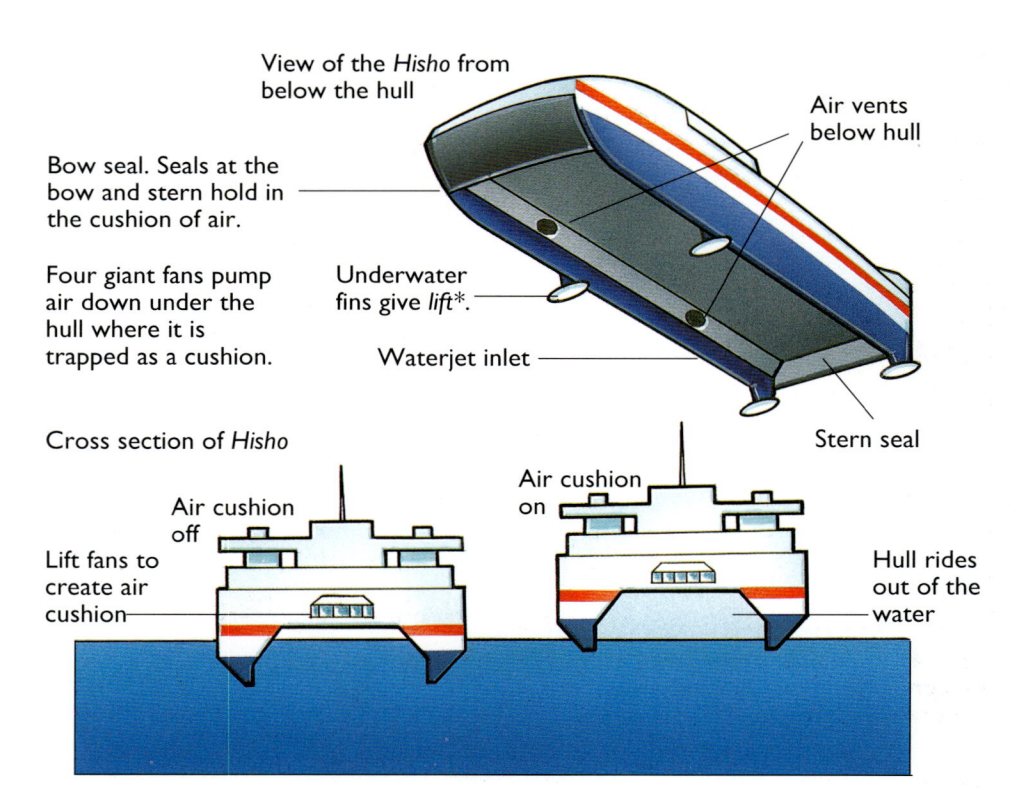

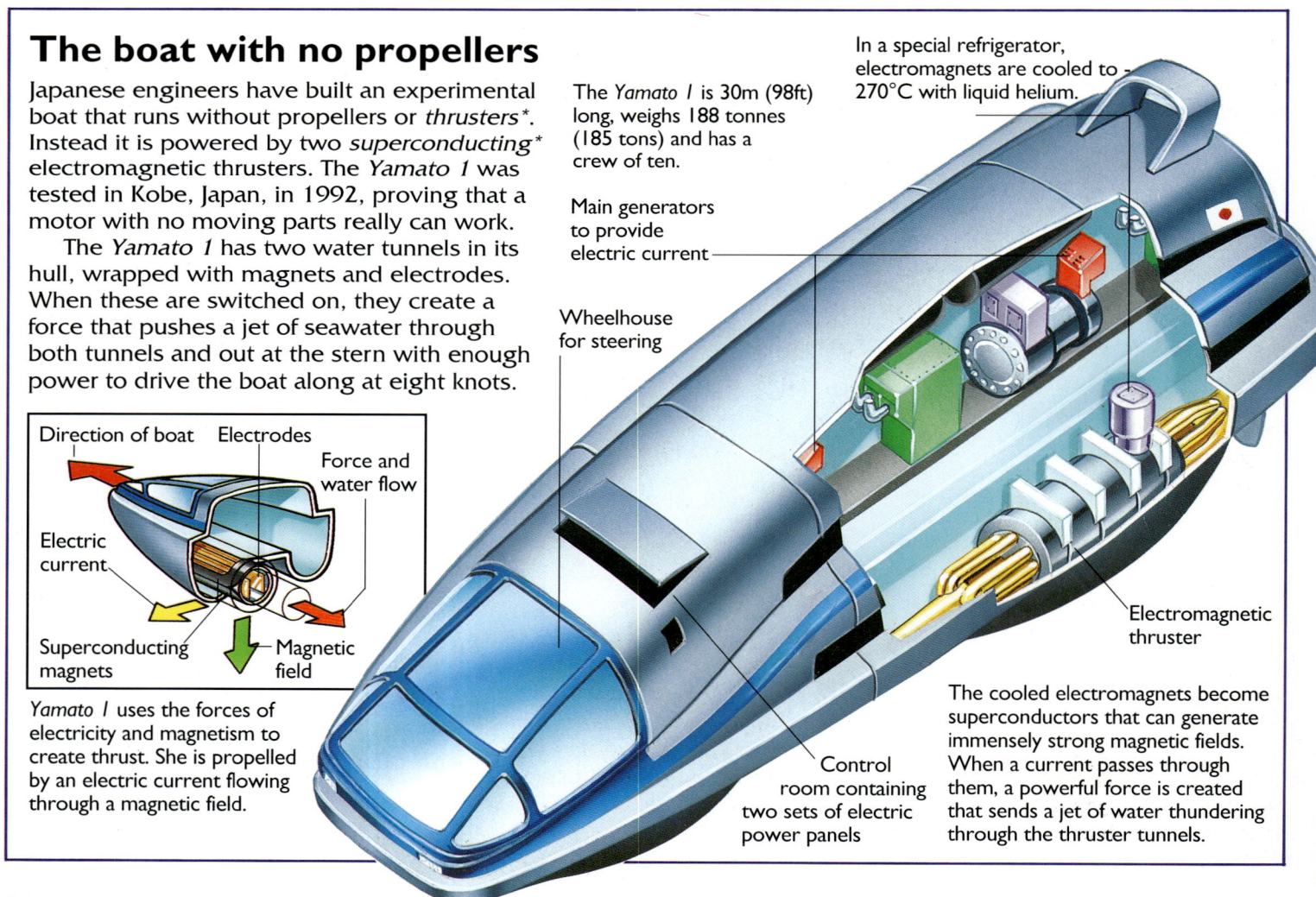

View of the *Hisho* from below the hull

Bow seal. Seals at the bow and stern hold in the cushion of air.

Four giant fans pump air down under the hull where it is trapped as a cushion.

Air vents below hull

Underwater fins give *lift**.

Waterjet inlet

Stern seal

Cross section of *Hisho*

Air cushion off

Lift fans to create air cushion

Air cushion on

Hull rides out of the water

The boat with no propellers

Japanese engineers have built an experimental boat that runs without propellers or *thrusters**. Instead it is powered by two *superconducting** electromagnetic thrusters. The *Yamato 1* was tested in Kobe, Japan, in 1992, proving that a motor with no moving parts really can work.

The *Yamato 1* has two water tunnels in its hull, wrapped with magnets and electrodes. When these are switched on, they create a force that pushes a jet of seawater through both tunnels and out at the stern with enough power to drive the boat along at eight knots.

The *Yamato 1* is 30m (98ft) long, weighs 188 tonnes (185 tons) and has a crew of ten.

Main generators to provide electric current

Wheelhouse for steering

In a special refrigerator, electromagnets are cooled to -270°C with liquid helium.

Control room containing two sets of electric power panels

Electromagnetic thruster

The cooled electromagnets become superconductors that can generate immensely strong magnetic fields. When a current passes through them, a powerful force is created that sends a jet of water thundering through the thruster tunnels.

Direction of boat — Electrodes — Force and water flow — Electric current — Superconducting magnets — Magnetic field

Yamato 1 uses the forces of electricity and magnetism to create thrust. She is propelled by an electric current flowing through a magnetic field.

Wave cutter

The knife-blade bows of fast catamaran ferries work so well that they may one day be used on cargo ships too. In the model below, the crew's quarters and the bridge are pushed up front, while as much deck space as possible is left free for freight. Up to 70 containers are stored out in the open, without hatches or covers. This saves weight and so makes the ship much faster.

This 40 knot wave-piercing freighter, designed by Incat, Sydney, Australia, is powered by four jet thrusters.

Flying boats

The most dramatic way to make boats faster is to lift them right out of the water altogether. One design that does this is the wing in ground-effect craft, or wingship.

The *Flarecraft L-325* rides smoothly on a cushion of air. Short stubby wings flying just above the surface of the water create a pocket of high air pressure (called ground effect) when they are moving at high speed. This will lift the 9.5m (31ft) craft into the air once it reaches 80km/h (50mph). But it cannot fly higher than 2m (6ft) above the waves, which is why it is registered with the US Coast Guard as a boat.

This *Flarecraft L-325* is a five-seat water taxi, which cruises at 120km/h (75mph).

Glossary

Ballast. Heavy weights, often lead or tanks of water, packed into the deepest parts of the hull or keel to give a boat better balance. Ballast stops a boat from rolling over in heavy winds and waves.

Bow. The narrow front end of a boat, pointed to cut cleanly into the water.

Bridge. The place from which ships are steered. Usually one of the highest places above deck with a good view.

Bulkheads. Walls and watertight doors that run from side to side inside the hull and divide it into watertight compartments.

Buoy. A bright float, anchored near ports, used for navigation or mooring.

Coxswain. The person who steers a boat. Also called a helmsman.

Deck. The floors of a ship.

Dinghy. Any small boat powered by sail, oars or outboard motor.

Drag. The force created by the action of water against the hull and propeller of a ship which slows it down. Ships with long, narrow hulls and pointed ends usually suffer less drag than those with wide hulls and blunt ends.

Driveshaft. The shaft that transmits power from the engine to the propeller in a ship. Also called the propeller shaft.

Funnel. The chimney of a ship which releases smoke and exhaust gases.

Horsepower. A measurement of a boat's engine power, equivalent to 746 watts.

Heat exchanger. An attachment to a ship's engine to pipe cold seawater past the hot water that cools an engine. The seawater draws off heat, without coming into direct contact with the engine.

Hull. The part of a ship which sits in the water.

Hydrofoil. A boat with underwater "wings" designed to generate lift. As speed increases, the hull is raised out of the water, so reducing drag.

Keel. The lowest structure of a ship's hull, running lengthways, upon which the framework of the hull is built.

Knots. The speed of a ship in water is measured in knots, or nautical miles per hour. One knot is 1.85km (1.15 miles).

Lift. The upward force created by wings.

Port. Facing toward the bow of a ship, its left-hand side is known as the port side.

Propeller. A rotating device, with two or more curved blades, that provides thrust for moving a ship forward. A propeller is attached to a shaft (usually at the back of the boat) that is turned by the engine.

Radar. A method of finding the position and speed of a distant ship or other object, by transmitting radio waves which are reflected back to the sender.

Rudder. A large blade at the back of a ship behind the propeller, for steering.

Stabilizers. Fins projecting from the sides of the hull, to help keep a ship steady.

Starboard. Facing toward the bow of a ship, the right side is known as starboard.

Stern. The back end of a boat, usually rounded so water flows smoothly past.

Superconducting. Having no electrical resistance. In metals this occurs when they are cooled to very low temperatures.

Thrust. The force which drives boats forward, provided by the turning action of the propellers which throws a powerful surge of water backward.

Thrusters. Extra propellers in the hull of a ship for moving sideways.

Turbine engines. High-speed engines that work like the jets that power planes.

Upthrust. The force pushing up on a boat when it is floating in water.

Wheelhouse. An enclosed platform from which a ship is steered. Also called the pilothouse or bridge.

Winches. Winding wheels for raising and lowering heavy anchors, or for hauling ropes to raise sails or tie up to a dock.

Index

ABS (anti-locking brakes), 22, 25, 29
acceleration, 25
accelerator, 27, 31
accelerometers, 58, 59
active safety features, 6
active suspension, 20
actuators, 39
Adolphus, Gustavus, 71
aerobatics, 49
aerodynamics, 16, 17, 29, 30, 31, 46, 47, 63
aerofoil, 63
aeroscreen, 19
ailerons, 36, 37, 48, 61
air bag, 6, 7
air brake, 37, 40, 51, 53
Airbus A300-600ST, 41
Airbus A340, 39, 40, 64
Airbus A3XX, 62
air corridors, 59
aircraft carrier, 52
air-cushioned vehicles (ACVs), 79
air filter, 4, 15
airflow, 67
air intake, 35, 37, 55, 60
airliners, 40, 41, 46, 47, 54, 62
air pressure, 14, 31, 35, 39, 63
airspeed indicator (ASI), 38, 39
air traffic control, 59
alloy, 42, 63
alternator, 11, 26, 31, 81
altimeter, 38, 39
altitude, 39, 54, 61
Alvin, 92
anchors, 77, 87
anti-G suit, 55
anti-roll bar, 21, 29
Antonov AN76, 43
aquaplanning, 23
arrester hook, 51
artificial horizon, 38, 58
Atlantic Companion, 86
automatic gearbox, 13, 28
autopilot, 59
avionics, 35, 45, 53
axle, 3, 20, 29

BAe/Aerospatiale Concorde, 43, 46, 47
BAe Hawk 200, 34, 35, 45
BAe Sea Harrier, 52, 53
balance weight, 10
ballast, 73, 95
ballast tanks, 93
ball bearings, 17
banked turn, 37, 38
barges, 88
battens, 77
battle, 69
Bell X-15, 62
Bentley 4 1/2 Litre, 18, 19
Benz Velo, 3
biplanes, 46, 49, 63
black box, 54
boathouses, 83
Boeing 747, 39, 47, 50, 62, 64
Boeing 757, 42
Boeing AWACS, 56
boilers, 67, 72, 73, 74
boom, 77
bow, 66, 67, 68, 75, 77, 79, 85, 87, 91, 95

brake horsepower (bhp), 11, 18, 19, 25, 28, 31
brake lights, 15, 26
brake pads, 22
brake parachutes, 51, 61
brakes, 51
bridge, 89, 91, 95
British Navy, 72
bucket seats, 8
bulkheads, 72, 91, 95
bunkers, 72
buoys, 78, 95
bypass duct, 42, 43

Cabins, 40, 66, 70, 74, 83, 85
cage, 7
caliper, 22
cam, 31
cambelt, see *timing belt*
camouflage, 61
camshaft, 11, 80
cannons, 70, 71
canoes, 66
canopy, 54, 55, 56, 60
captain, 70, 79, 91
cargo, 74
cargo hold, 41, 62
cargo planes, 35
cargo ships, 82, 86, 89, 95
catalytic converter, 25, 29
catamarans, 84, 85, 94
catamaran ferries, 95
catapult, 51
Cessna 150, 48
chassis, 4, 31
chocks, 52
clutch, 12, 27, 31
coal, 66, 67, 72
cockpit, 14, 35, 36, 37, 38, 39, 48, 49, 52, 57, 58, 62
co-driver, 8, 9
cogs, 12, 13
coil, 26, 27
combustion chamber, 43, 63
combustion cycle, 10
compass, 58
composites, 45, 61
compression chamber, 42
compressor, 53, 63
compressor blades, 42, 43
Computer Aided Design (CAD), 4, 45
computers, 86, 90, 91
Concorde, 46, 47
connecting rod, 10
contact points, 27
container ports, 87
containers, 95
container ships, 86-87
control cables, 36, 37
control column, 37, 38
control surfaces, 36, 37, 38, 46, 61, 63
coxswain, 82, 83, 95
crank, 31
crankshaft, 10, 11, 27
crash test dummy, 6
cross ply tyres, 31
cruiseships, 90, 91
crumple path, 7
crumple zone, 7
Cuban Eight, 49
Curtiss Hawk, 44
cylinder, 10, 11, 20, 22, 25, 26, 38, 42

Damper, 20, 28, 29
Dassault Rafale, 54
dead reckoning, 58
deck, 67, 68, 69, 70, 71, 74, 75, 76, 87, 90, 95
deck winch, 72
De Havilland Leopard Moth, 38
Delaunay-Belleville F6, 2, 3
delta wings, 35
depth finder, 76
DHC-2 Beaver, 50
diesel engines, 76, 80, 82, 86
differential, 12, 13, 31
dinghies, 77, 95
dipped headlights, 27
dipstick, 4
disc brake, 5, 15, 22
distributor, 26, 27, 31
downforce, 14, 25
drag, 36, 37, 46, 51, 55, 85, 95
drag coefficient, 16
dredgers, 88
driveshaft, 12, 13, 80, 95
drogue basket, 47
drogue chute, 55
drum brake, 5, 22, 25
duralumin, 44, 63

ECM pod, 57
EFA EuroFighter, 54
Egyptians, Ancient, 66
ejection seat, 45, 52, 54, 55
electric engines, 11, 31
elevators, 36, 37, 48, 50, 51, 61
engine room, 68
engines, 34, 40, 42, 43, 44, 50, 51, 67, 72, 78, 80-81, 84
exhaust gases, 61
exhaust nozzles, 52, 53, 57
exhaust pipe, 5, 8, 15, 29
exhaust valve, 10

Facetting, 60, 61
Falcon 900, 50
fan, 4, 24
fan blades, 40
ferries, 78-79, 81, 82
fighting ship, 71
fire extinguisher, 9
firefighting tug, 89
fishing boats, 81
flaps, 79, 82
Flarecraft L-325, 95
floating, 66, 68, 73
flying boats, 95
fog lamp, 26
Ford Escort RS Cosworth, 8, 9
Ford Trimotor, 40
foremast, 68, 77
Formula I boats, 84
four forces of flight, 36, 63
four wheel drive (4WD), 12, 24, 28
freight, 67, 74, 86, 95
friction, 16, 17, 31, 85
fuel, 67, 80, 83
fuel injection system, 10
fuel management systems, 47
fuel pipe, 10
fuel tank, 34, 45, 47, 52, 56, 57, 61, 62, 83
funnel, 67, 72, 73, 88, 95
furnace, 72, 73

Galley, 46, 63, 70
gas turbine engine, 42, 43, 46, 62
gearbox, 12, 13
gears, 12, 13, 27, 31
gear stick, 13, 27
General Dynamics F16, 39, 49
gliders, 38, 48
gravity, 36, 55, 63
Greeks, Ancient, 68, 69
gudgeon pin, 11
gyroscope, 38, 58, 59

Handbrake, 31
handling, 20
hard point, 53, 57, 61
Hawker Hunter, 46
headlight, 3, 4, 9, 19, 26, 27
head-up displays (HUDs), 30
heat exchanger, 95
helmsman, 76, 95
high tension lead, 27, 31
Hisho, 94
H.M.S. Gannet, 72, 73
Honda Dream, 31
horizon, 58
horsepower, 82, 95
HOTAS, 63
hull, 67, 68, 69, 71, 72, 73, 74, 76, 77, 82, 83, 84, 85, 87, 94, 95
hydraulic cylinder, 20, 22
hydraulics, 20, 31, 38, 39, 45, 49
hydraulic system, 22
hydrofoils, 85, 95

Idler gear, 13
ignition, 25, 26, 27, 31
inboard engines, 81
indicator, 26
inlet manifold, 10
inlet valve, 10
input gear, 13
input shaft, 12, 13
INS, 63
intake fan, 42, 43, 52
internal combustion (I.C.) engine, 3, 10, 11
IZA electric car, 11

J250 craft, 85
jet engine, see *gas turbine engine*

Keel, 71, 77, 83, 95
Kevlar, 45
knots, 68, 73, 83, 95

Laminated glass, 5
layshaft, 13
leading edge, 35
Le Mans race, 18, 24
lifeboats, 82-83, 89
life jackets, 54
life rafts, 54, 55
lift, 35, 36, 37, 46, 50, 51, 67, 84, 85, 94, 95
light aircraft, 36, 38
lightweight boats, 70

Lockheed C17, 56
Lockheed F117A, 51, 60, 61
Lotus 107, 14, 15
lubricant, 11, 15, 17, 31
LVG CVI bomber, 44, 45

M
Magnetic field, 59, 63
main sheet, 77
maintenance, 48
marshals, 9
masts, 68, 70, 76, 77, 78, 79
military submarines, 93
Mille Miglia race, 18
mini submersibles, 93
monocoque, 4, 44
monoplane, 46, 63
monoshock, 14
Monte Carlo race, 24
muffler, see *silencer*

N
Naval Experimental Manned Observatory (NEMO), 93
navigation, 38, 78, 79, 82, 89
Norbert, 92
Northrop B2, 60
Northrop T38 Talon, 48
nosewheel, 50

O
Oil filter, 11
oil tankers, 82
Olympias, 68
Olympus 593 engines, 43, 47
Optica, 38
outboard engine, 80, 84
output gear, 13
output shaft, 13

P
Packets, 74
paddlewheels, 74, 75
Panavia Tornado, 56
panhard rod, 21
parachute, 54
passenger cell, 7
passenger configuration, 40, 41
passive safety features, 6
Peugeot 905B-EV11, 18
pinion, 21
piston, 10, 11, 20, 26, 38, 42
piston engines, 42
pitch, 36
pitot tube, 39, 60
Pitts Special, 49
plies, 23
pneumatic, 29, 31
pod, 15
pollution, 11, 23, 25, 29, 30
Porsche 911 Carrera 4, 24, 25
Porsche 959, 25
ports, 86, 87
port side, 67, 95
Pratt and Whitney F100-PW-100 engine, 47
pressurization, 40, 63
pre-tensioners, 6

Pride of Calais, 78
production line, 4
propellers, 43, 67, 72, 74, 76, 79, 80, 83, 86, 87, 88, 90, 94, 95

R
Racing boats, 84-85
 catamarans, 84, 94
 hydroplanes, 84
 monohulls, 84
racing harness, 8
rack (gear), 21
radar, 35, 38, 52, 53, 54, 56, 57, 58, 59, 60, 61, 78, 79, 82, 83, 89, 95
radar absorbant material (RAM), 60
radar cross section (RCS), 60
radial tyres, 23, 31
radiator, 4, 19
radio, 54, 58, 76, 82, 91
rally cars, 8, 9
ram pressure, 39
range, 40, 63
Range Rover LSE, 28, 29
reaction jets, 53
rear view mirror, 5, 14
reconnaissance, 56
Red Arrows, 35
ride height, 29
riverboats, 74-75
Rob't. E. Lee, 74-75
rocket engines, 62, 63
roll, 36
roll cage, 8, 9
Rolls Royce Pegasus, 52
Rolls Royce RB211-535, 42, 43
Rolls Royce Silver Ghost, 3
roof vent, 9
rotor arm, 27
rowers, 68, 69
 thalamites, 68
 thranites, 68
 zygites, 68
Royal National Lifeboat Institution (RNLI), 82
rudder, 36, 37, 45, 61, 67, 75, 76, 77, 79, 95

S
SA35 engine, 18
sailplanes, see *gliders*
sails, 66, 67, 68, 69, 70, 71, 72, 77
 foresail, 76
 genoa, 77
 jib, 76, 77
 mainsail, 77
 mizzensail, 76
 spinnaker, 77
satellite, 76, 82
screw, see *propeller*
seat belt, 5, 6
servos, 59
shafts, 12, 13
side fairing, 15
side impact bar, 5, 7, 24
side keel, 83

side-wheelers, 75
silencer, 29
slats, 37, 50, 51
slick tyres, 15
Slingsby T67 Firefly, 36, 37, 45
solenoid, 26
Space Shuttle, 63
spark plug, 10, 11, 25, 26, 27
Special Stages, 9
spoilers, see *air brakes*
spring, 20, 21, 28, 29
SR71 Blackbird, 61
stabilizers, 79, 90, 95
stacking, 59
stalling, 37, 49, 63
stall turn, 49
stall warning system, 37
starboard side, 67, 95
starter motor, 26, 80, 81
starting handle, 3
static air pressure, 39
stealth, 60, 61
steam engines, 66, 72, 74
steamships, 72-73
steam tractor, 2
steering, 66, 67, 79
steering column, 6, 19, 21
steering oar, see *rudder*
stern, 67, 68, 69, 74, 75, 77, 79, 87, 95
streamlined, 16, 30, 31, 46. 50, 63, 84, 85
struts, 46
studded tyres, 8
submarines, 92, 93
submarining, 6
submersibles, 92-93
suction, 80
Sukhoi Su26, 44
Sukhoi Su35, 61
sump, 11
Sun Princess, 90
supercharger, 19
superconducting, 95
Supermarine Spitfire, 46
Surfury, 84
suspension, 5, 20, 21, 28, 29, 30
Swan 55, 76
swing wings, 56

T
Tachometer (rev. counter), 27, 31
tacking, 77
tailspin, 49
tailwheel, 50
telemetrics, 15
thermals, 48
throttle, 39, 50
thrust, 36, 42, 43, 51, 53, 62, 80, 95
thrusters, 79, 86, 87, 92, 94, 95
thrust reversers, 47, 51
tiller, 69, 77
timing belt, 11
Titanic, 92
titanium, 93
torque, 12, 31
trailing edge, 35

trainers, 48
transmission, 12, 13
tread, 15, 23
triremes, 68-69
tug boats, 66, 81, 87, 88-89
tug planes, 48
tuning, 31
Tupolev Tu204, 42
turbine engines, 42, 62, 63, 81, 95
turbofan, 42
turbojet, 43, 47
turboprop, 43
turning circle, 21

U
Undertread, 23
universal joint, 12
upthrust, 66, 95

V
Valve, 10, 11, 20, 31
valve cam, 10, 11
Vasa, 70, 71
Vasa Museum, 71
vectored thrust, 52
Victory boat, 85
VIFFing, 53
virtual reality, 31
Voith-Schneider propellers, 88
Volkswagen Beetle, 32
Volkswagen Golf, 4, 5
volume, 18, 31
VSTOL, 52, 53

W
Warships, 68, 70, 81
water jacket, 10
water jets, 80
water pump, 11
water taxi, 95
water tunnels, 94
wheel arch, 16, 28
wheelhouse, 66, 86, 94, 95
wheel spin, 31
winches, 77, 88, 95
wind direction, 67, 68, 77
windshield, 3, 5, 16, 19
wind tunnel, 4, 17, 47
wings, 14, 15, 25, 35
wing mirror, 4, 16
wing ribs, 44, 45
wingspan, 35, 53, 57
wingtips, 34, 35, 53
wood, 73
Wright Brothers, 34, 64

Y
Yachts, 76-77, 81
 ketches, 76
 schooners, 76
 sloops, 76
 yawls, 76
Yamato 1, 94
yaw, 36, 37, 52

Z
ZETEC engine, 10, 11

First published in 1996 by Usborne Publishing, 83-85 Saffron Hill, London EC1N 8RT, England.
All rights reserved. No part of this publication may be reproduced, stored in a retrieval system, or transmitted in any form, or by any means, electronic, mechanical, photocopying, recording or otherwise, without the prior permission of the publisher.
Copyright © 1996 Usborne Publishing Ltd. The name Usborne and the device are Trade Marks of Usborne Publishing Ltd.
Printed in Italy.